BRIDGING
THE GAPS

An Anthology on Nuclear Cold Fusion

Compiled and edited by

Randolph R. Davis

WESTBOW
PRESS®
A DIVISION OF THOMAS NELSON
& ZONDERVAN

WestBow Press books may be ordered through booksellers or by contacting:

WestBow Press
A Division of Thomas Nelson & Zondervan
1663 Liberty Drive
Bloomington, IN 47403
www.westbowpress.com
844-714-3454

Cover with Star of Bethlehem designed by the author.
Ocean scene from ©Getty Images.

ISBN: 978-1-6642-3416-1 (sc)
ISBN: 978-1-6642-3415-4 (hc)
ISBN: 978-1-6642-3417-8 (e)

Library of Congress Control Number: 2021909609

Print information available on the last page.

WestBow Press rev. date: 6/3/2021

In Memory of Melba G. Davis, Walter C. Davis, and Anne B. (Omie) Davis for their constant godly inspiration and support.

CONTENTS

Preface...xi

1. Introduction.. 1

2. Cold Fusion Energy Production 11
   Chemical Reactions 12
   Nuclear Reactions 15
   Standard Fusion Reactions................................. 17
   Energy and Products from Cold Fusion............. 19
   Transmutation Reactions 26

3. Phonons and Their Role in Cold Fusion.................... 29
   Macroscopic View 29
   Microscopic View 31
   Cold Fusion Theory 33

4. Nuclear Radiation from (p, d) Fusion 40
   Radiation Measurement Issues 41
   Utility of (p, d) Fusion 42
   Gamma Ray Attenuation 43

Radiation Safety... 46
Radiation Safety in 200 kW Generator Design.... 47
Radiation Exposure.. 51
Absorbed Dose.. 52
Equivalent Dose.. 52
Comparison to Other Radiation Sources 54
Comparison to NERVA and Cobalt-60
       Heat Sources 56

5. Scale-up and Industrialization 59
Problems with Hot Fusion...................................... 61
Problems with Fission Power Plants 64
US Government Participation 64
Supporting Patent Knowledge Base..................... 65

6. Robust Cathode Design.. 73
Technical Background 74
Calculation of Energy Limitations....................... 76

7. Generator Design 80
Supporting Missions to Outer Space 81
Reaction Chamber 81
Heat Exchanger/Boiler... 84
Gas Handling System 86
Electronic Control Subsystem............................... 88

8. Steps for Advanced Development and
Demonstration... 90
Estimated Program Cost.. 92
Program Activities.. 92
Test Planning and Preparation 93

Laboratory Testing in a Relevant Environment ... 95
Prototype Demonstration 97
Validation and Verification 99
Preparation for Production 100

9. Demonstration Experiment 102
Design of a 10 kW Demonstration 102
Construction 104
Assembly and Operation 105

10. Benefits of Cold Fusion Technology 108
Deuterium Cost 109
Community-Based Power Plants 110
Future Use Cost Analysis 112
Conclusions 113

References 117

Appendix 125

PREFACE

The earth needs a revolutionary transformation in the production of energy. Oceans are eroding land at an alarming rate. Due to rapid acidification of oceans from dissolved carbon dioxide, the entire chain of sea organisms is at risk. Expanding populations and rapid industrialization have fostered massive reliance on and competition for fossil fuel. The result is cities choked with crippling pollution. In other countries with enormous growing populations, millions are without even minimal amounts of clean water, which would be easily supplied if suitable energy sources were available. Populations have been increasingly reliant on fossil fuels, which are known to pollute our planet and also will be depleted or too expensive to use over the long term. The fracking process to increase gas supplies is severely detrimental to the environment and harmful to our health. Disasters have shut down the nuclear power industry and forced an increased reliance on fossil fuels. Many scientists warn of worldwide catastrophic effects of climate change from global dependence on fossil fuels, and the costs and competition for energy are sources of international conflicts, fomenting warfare and human destruction.

Conventional nuclear power was a hopeful alternative in the last half century. Now there is evidence of the downside— possible catastrophic accidents, unsolved problems of high-level radioactive waste requiring secure storage for thousands of years, weapons proliferation, and susceptibility to natural disasters, terrorism, and war. Other alternatives have been pursued—solar power, wind power, and more. One of the difficulties in the success of these approaches is the intermittent nature of the source, and no suitable energy storage technology has emerged that can overcome this disadvantage. Nuclear fission power plants of any size are no longer considered to be acceptable methods to supply energy due to the short- and long-term nuclear radiation that they produce. Plans to build nuclear fission plants have been cancelled in many countries. The long-promised hot fusion technology has not been able to reach the point of breakeven where a net amount of energy is produced. Such machines are many decades from actual application and, if successful, will likely be prohibitively expensive and economically viable only for supplying energy over large areas.

For the last thirty years, scientists have been working to discover how to apply cold fusion technology to produce useful amounts of energy. Research by a small group of professional and amateur scientists in Northern Virginia, as well as others worldwide, indicates that this may be possible. This technology could possibly address all of the above problems and could do so at reasonable cost, in safety, without hazardous waste, and without pollution or climate damage. This is possible because of the enormous energy yield produced by nuclear reactions of a type that are not the same as those in conventional nuclear power plants that have produced hazardous radiation and by-products from fission. Cold fusion involves nuclear reactions that occur *at low energies* and has been studied in an array of configurations with many differing materials and under different operating conditions. The most amazing facet of this new technology is that its typical fuel, deuterium, is derived from water, and *this fuel source could supply the entire energy demands of the earth for thousands of years.* The process, additionally, has no application in weaponry, and the universal access to fuel could prevent political and military rivalries that compete for access to energy.

INTRODUCTION

THE ANNOUNCEMENT[1] OF COLD FUSION IN 1989 BY Martin Fleischmann and Stanley Pons at the University of Utah, and of related work by Stephen Jones at Brigham Young University, was an early revelation of future energy technology that may now become a blessing to civilization. Could it have been an omen of things to come ... civilization's need for a new source of electric power? The world today is in grave danger from global warming and climate change just two hundred years since the beginning of the Industrial Revolution. While industrial progress has significantly enhanced humanity, the *devastation caused by global warming* indicates that we are at a critical junction of human history.

The criticism surrounding cold fusion in the first few years after its announcement in 1989 largely debunked this area of science and technology in the eyes of many scientists and laypersons alike. If the experiments had produced high-energy neutrons from deuterium-deuterium (d, d) fusion, then capture of the neutrons by protons (i.e., hydrogen) in the calorimeter's surrounding H_2O cooling water would have produced detectable 2.22 MeV gamma radiation, which was not observed. Another argument by critics was that, if the cold fusion experiments actually worked, even producing only a watt of power, then nearby scientists would have received a lethal dose of radiation, but they seemed to have no ill health effects. A third criticism was that, if cold fusion were real, then products of the reactions, such as helium and tritium, would have been able to be detected in quantities that correspond to the measured excess heat. Critics also asserted that the experiments would need to be repeatable, or reproducible, for acceptance of cold fusion by the larger scientific community.

The first two arguments now seem to overlook some aspects of nuclear physics related to design of the experiments, or there is little evidence that they were discussed at the time. Heavy water (D_2O) used in cold fusion is also a neutron moderator for some other applications. Neutrons, if they were produced by (d, d) fusion, might have been slowed down, or moderated, by heavy water near the center of the cold fusion experiments. The lower-velocity neutrons may not have been able to traverse the internal container to reach water in the calorimeter. Neutron absorption calculators could have been used to determine if the neutrons would be attenuated through the heavy water and calorimeter walls, but this doesn't seem to have been addressed. The criticism that experimenters would have received a potentially lethal radiation dose does not seem to consider the limited time that researchers were near their experiments or the possibility that neutrons could have been moderated by the heavy water and calorimeter. Little interest was shown for alternative methods that might have been used to detect radiation.

What was not foreseen at the time was that, instead of just simply ending up in the annals of scientific history, cold fusion would continue to be studied worldwide by professional and amateur scientists over the next thirty years and to the present time. During this time, scientists have verified that fusion-type nuclear reactions can be made to occur without a high voltage or temperatures required for hot fusion. A product of the cold fusion reactions, helium, has been shown to correspond in quantity to the excess heat measured in the experiments. Nuclear particles, including neutrons, have been detected in some experiments.[2]

These advancements have been possible through the application of accurate scientific instrumentation and measurement methods and the ability to discuss results easily with the international scientific community. The acronyms LENR (low-energy nuclear reactions) and CANR (chemically-assisted nuclear reactions) are sometimes used for related work. Technical reports describing this progress are available on the internet in a publication library provided by the International Society for Condensed Matter Nuclear Science, through a LENR-CANR website, and on other websites where cold fusion is discussed.[3, 4] Annual international conferences and other research meetings have been held over the last thirty years to discuss cold fusion. Related reports and presentations are included in technical journals and discussed on the internet.

Progress was discussed on May 5, 1993, by the Congressional Energy Subcommittee, US Committee on Science, Space, and Technology. Department of Energy (DOE) representatives at the hearing reviewed the status of magnetic confinement, or hot fusion. A scientist from industry pointed out that hot fusion funding excluded virtually all other research on new ideas and systems. Another scientist from industry described research on a chemical process tangentially related to cold fusion. Concerns were expressed that absence of an official US policy on cold fusion prevented a much-needed intellectual property infrastructure and that US patents were not being issued in this area of technology. Intellectual property regarding cold fusion, however, has been protected since 1994 through the patent-approval process.

The DOE held a meeting some eleven years later on August 23–24, 2004, at the request of members of the scientific community, to review progress in cold fusion. Information about the review is discussed on the internet.[5] Scientists who requested the review provided an in-depth paper, "New Physical Effects in Metal Deuterides," (August 1, 2004) on progress until that time, along with 137 supporting background papers. Eighteen participants in the review provided comments. About half of these reviewers in 2004 indicated that cold fusion can be produced and that excess power was demonstrated in many of the experiments discussed in the review. Improved measurement of heat produced was needed to demonstrate reproducibility.

Nuclear physics is a complex technical field of study. Many theories have been developed to understand its unseen world of small nuclear particles and electromagnetic waves since Ernest Rutherford's alpha particle scattering experiment in 1911. This has led to complex mathematical methods and nomenclature to describe its many related properties and effects. Thus, steps in understanding nuclear physics and cold fusion can be accompanied by many questions and have led some scientists either to question their knowledge of the past or be amazed at the technical progress and results. One highly respected scientist working on cold fusion recently said, "It's nice to dream, isn't it?" while another has concluded that *the implications are as profound as was the discovery of fire.*

The need for reproducibility has been discussed in many meetings of the cold fusion community. A recent report, "Evidence of Operability and Utility from Low Energy Nuclear Reaction Experiments," indicates that operability is demonstrated when a cold fusion or LENR device produces heat or nuclear reaction products.[6] Examples of experiments that produced heat and helium are provided on pages 10–19 and 28–36 of the report. Utility of an operable device is demonstrated when its design is subsequently used in the design of another "operable" cold fusion or

LENR device. Reproducibility can be demonstrated with the same or different devices.

One of the cold fusion teams of professional and amateur scientists in Northern Virginia, USA, has worked independently since the mid-1990s to understand this field of science. Their work has mainly involved review, analysis, and discussion of the considerable amount of technical information in cold fusion literature. From this, they developed the design of a cold fusion energy generator to bridge the gaps between experimental laboratory cold fusion systems and a commercially useful device. This book attempts to capture this understanding, with the hope that advanced research and development (R&D) companies will be able to apply the information successfully in further steps of cold fusion generator development.

Most cold fusion experiments have involved liquid electrolysis where an anode and cathode are immersed in a liquid electrolyte (heavy water, or D_2O). Some experiments have investigated cold fusion of deuterium gas in a metal matrix, without using a liquid electrolyte. Liquids produce problems, such as boiling and evaporation of the liquid, buildup of contaminants, and limited operating temperature. Scale-up from liquid electrolyte experiments to industrial systems would be difficult to impossible due to these types of problems. The Northern Virginia team, therefore, focused upon designing a gas or *gaseous* type of system. It includes a central reaction chamber that is surrounded by a heat exchanger/boiler. The reaction chamber is designed to contain high gas temperature and pressure during operation and to be opened and closed for maintenance. The reaction chamber contains a large-surface-area cathode. Reaction material in the cathode is made by high-pressure, metal powder consolidation. A thermal gradient combined with gas pressure and strong electric field are used to load the cathode. The design also includes a gas-handling system consisting of four independent gas manifolds.

In this process, the team identified a number of analytical issues that appear to have been overlooked by other researchers or need to be better understood. These issues were noted when information for analysis was missing from the literature, or information provided was so novel that it could not be sufficiently understood for useful application. These are some of the types of gaps that this book endeavors to bridge, and they should be studied further in future cold fusion generator development programs. Examples are the following:

- Although conditions for cold fusion must be completely different from those of hot fusion, some researchers nonetheless speak of energy needed to produce cold fusion as if it were directly related to a high accelerating voltage (e.g., 100 keV) or high temperatures required in hot fusion. Since high temperatures or acceleration voltage are not needed for cold fusion, this can divert attention from understanding the more important cold fusion parameters.

- Sodium iodide (NaI) scintillator detectors are normally used as a method to measure gamma radiation but can be used for neutron detection. The important parameters for neutron detection (e.g., configuration and efficiency) needed for analysis were inadequately addressed in the cold fusion literature.

- Gamma radiation from liquid electrolysis cold fusion experiments might be emitted in preferred directions versus isotropically, or equally in all directions (expected in hot fusion). Little to no discussion was found in the cold fusion literature concerning preferred directions for gamma ray emission. As a consequence, the best detector locations might not have been chosen for making radiation measurements.

- High-energy gamma radiation was not observed from the experiments, but little attention was given to the possibility of internal conversion and pair production as alternative processes for an excited nucleus (e.g., helium-4) to reach its lower-energy ground state by emitting high-energy electrons. Gamma ray emission was emphasized as if it were the only way for an excited nucleus to lose energy and reach its ground state. As a result, researchers devoted significant time and resources developing alternative theories to explain the reason gamma radiation is not detected.

- In a cold fusion environment, proton-deuteron (p, d) fusion should theoretically have a higher probability of occurring than deuteron-deuteron (d, d) fusion.

Nonetheless, only a small amount of information could be found in the cold fusion literature about experiments to use hydrogen in combination with deuterium. Some researchers surmised that instead of (p, d) fusion to produce helium-3, the proton might combine with an electron, forming a neutron, which might then enter the deuteron to form tritium, rather than helium-3. That is, however, not expected to occur with high probability.

- Energy generators based upon cold fusion should be able to operate for much longer periods than generators based on transmutation. Either process might provide similar amounts of energy. Little discussion, however, could be found in the literature to justify LENR transmutation experiments, instead of focusing on cold fusion.

This book also provides additional information developed by the team to design its cold fusion generator. Supporting technical references, including internet addresses, are indicated in a separate reference section.

Examples of conversion factors used in the text include the following:

1 million electron volts (MeV) = 1.6×10^{-13} joule
1 watt (w) = 1 joule/second
1 Angstrom (A) = 10^{-10} meter
1 atomic mass unit (amu) = 931.5 MeV
1 barn = 10^{-24} cm^2

The team's design is anticipated to have near-term applications in laboratory testing of cold fusion concepts, gas electrolysis, cooling and heat transfer to promote cathode loading, energy conversion to provide useful power, and cooling methods to enable long periods of operation. The ultimate objective is to contribute to development of a commercially useful cold fusion system as the prime power source for community-based power plants. It is hoped that this information will provide technical clarity for managers and scientists in R&D companies taking the next bold steps in cold fusion technology development.

2

COLD FUSION ENERGY PRODUCTION

NUCLEAR FUSION IS A PROCESS IN WHICH NUCLEI join, or fuse, to form a product nucleus. Because the small nuclei are positively charged, they repel each other, and only nuclei that move quickly, with enough kinetic energy, actually fuse. Energy is produced due to the difference in mass of the product nucleus (or nuclei) and the mass of the initial nuclei. The missing mass is converted to energy. In hot fusion, high-speed nuclei are created through particle accelerators or heating gas to extremely high temperatures. In cold fusion, conditions for the reaction cause fusion at a much lower temperature than previously thought possible.

In the process of performing cold fusion experiments, scientists have determined that more energy (heat) can be produced than is possible with chemical reactions. Chemical reactions involve electron volts (eVs) of energy per reaction, while nuclear reactions typically involve millions of electron volts (MeVs) of energy per reaction. One electron volt = 1.6×10^{-19} joule, but one MeV = 1.6×10^{-13} joule, which is a million times larger. In addition, scientists have observed that various types of atoms can be produced and that these might be the result of nuclear fusion, nuclear fission, nuclear transmutation, or a combination of these three types of reactions.

Chemical Reactions

It is first important, due to the relatively small amount of energy produced in oil, gas, and coal power plants, to discuss how energy (heat) is produced in chemical reactions. Atoms bond together to form molecules because, in doing so, they attain lower energies than they possessed as individual atoms. That is, the bonded atoms or molecules have an energy that is less than the sum of energies of the initial atoms. A quantity of energy is released that is equal to the difference between the energies of the atoms bonded together to form molecules and the energies of the initial atoms or molecules. The energy is usually released as heat. When atoms combine to make molecules, energy is always emitted, and the compound has a lower overall energy. The internet provides an abundant amount of information on chemical reactions.[7, 8]

During a chemical reaction, atoms are rearranged and bonds are broken within the reactant molecules as new bonds are formed to produce product molecules. This involves breaking of chemical bonds between atoms of reactant molecules and forming new chemical bonds between atoms of product molecules. Bond energy is defined as the amount of energy it takes to break one mole (6.02×10^{23}) of bonds in the gas phase. Energy always has to be added to break a chemical bond. Bonding always releases energy. If more energy is released during bond formation (of the products) than bond breaking (of the reactants), then the overall reaction is exothermic. This can be represented by the following diagram:

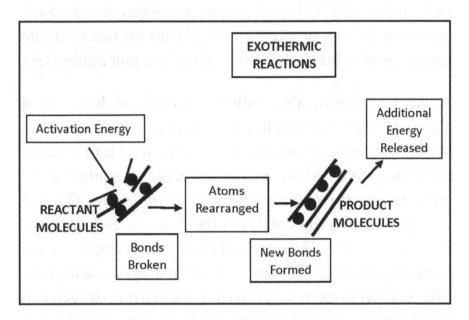

Activation energy is defined as the minimum amount of energy needed to activate atoms or molecules to a condition in which they can undergo chemical transformation. It is sometimes much less than the bond energy of reactants and can even be close to zero for some reactions.

Energy is required to break bonds, and energy is released when bonds are formed. The numerical value of the bond energy is the same whether a bond is being broken or formed. The change in energy is positive (+) when bonds are broken and negative (-) when bonds are being formed. The minus sign is important to indicate, as this signifies heat energy is released when bonds are formed. The total energy change of the reaction (or "enthalpy of the reaction") is the energy it takes to break the bonds of reactants plus the energy that it takes to produce new bonds. It is generally indicated as ΔH. This is the same as subtracting the total amount of energy produced as bonds are formed from the energy used to break the bonds of the reactant molecules.

A covalent bond, also called a molecular bond, is a chemical bond that involves the sharing of electron pairs between atoms. In a covalent bond, atoms with the same electronegativity share electrons because neither atom preferentially attracts or repels the shared electrons. The best examples of covalent bonds are the diatomic elements like H_2, N_2, O_2, and F_2. Water (H_2O) is another example, as it is formed by sharing electrons between hydrogen and oxygen. The reaction to form water from hydrogen and oxygen can be represented in the following equation. Activation energy is provided by heat from a flame contacting the gas.

$$2\ H_2 + O_2 \rightarrow 2\ H_2O \qquad \Delta H = -481\ kJ.$$
$$\text{(flame)}$$

The H-H bond in hydrogen gas has an energy of 436 kilojoules per mole (kJ/mol). The O-O bond in oxygen gas has an energy of 499 kJ/mol. Therefore, the energy needed to break the bonds of two moles hydrogen and one mole of oxygen is (2 x 436) + 499 = +1371 kJ. The O-H bond in water has an energy of 463 kJ/mole and a molecule of water has two O-H bonds. The energy needed to form both bonds in two moles of water molecules is 2 x 2 ×463 = -1852 kJ. The total energy change during this chemical reaction is then +1371 + (-1852) = -481 kJ.

A watt of power is defined as one joule of energy per second (1 Watt = 1 J/sec). One kilowatt (kW) is 1 kJ/sec (1000 J/sec). Thus, 481 kW (481 kJ/sec) of power can be produced by burning two moles of hydrogen and one mole of oxygen each second. In comparison, a house requires about 4–5 kW of power for heating, lighting, and so on.

Nuclear Reactions

Nuclear reactions are quite different in that they involve interactions with the very small, internal nuclei within atoms. Atomic bonds are not considered. The nuclei within atoms consist of neutrons and protons and are positively charged due to the protons they contain. Nuclear reactions also involve nuclear particles (e.g., alpha particles, beta particles, neutrons) produced from these internal nuclei. The internet provides valuable information on how nuclear reactions work.[9, 10]

The probability of an effective nuclear interaction depends on the size (cross section) of the interacting particles and the velocity (or kinetic energy) with which the particles interact. Due to their small size, it is a statistical process. The interacting nuclei or particles have to be on nearly exact but opposite trajectories so that they will collide, or one of the particles may be stationary. A positively charged particle, such as a proton, a deuteron, or an alpha particle, can only interact effectively with a positive nucleus in an atom if its velocity is also sufficient to overcome the coulomb barrier produced by its proton(s) and the protons in the nucleus. In fusion, two positively charged nuclei, such as a proton and deuteron or two deuterons, can interact and combine if their velocities are sufficient to overcome the coulomb barrier produced by their protons. The height of the coulomb barrier represents an effective threshold energy for nuclear reactions involving charged particles. A neutron does not contain a positive charge and can more easily interact with nuclei in atoms.

The probability of a nuclear reaction is described by the manner in which the reaction's cross section varies as a function of kinetic energy of the interacting particles. Cross sections have previously been determined from laboratory measurements and are in units that can be compared to a typical geometric cross section area for a nucleus, which is about 10^{-24} cm^2 (this unit of measure is called a "barn"). The cross section for neutron-induced reactions increases with decreasing velocity or kinetic energy because the likelihood that a neutron can be captured depends upon the amount of time it spends near a particular nucleus. The cross section for positively charged particles increases with increasing kinetic energy because of the presence of the coulomb barrier.

Standard Fusion Reactions

As indicated above, some in the scientific community believe that cold fusion is possible because they were able to determine from their experiments that MeVs of energy were produced in the form of heat. In addition, nuclear reaction products, such as helium, have been produced by some experiments. The helium has been correlated with heat produced. It is assumed, therefore, that energy from cold fusion can be produced by the same types of nuclear reactions (fusion, fission, and transmutation) historically observed for standard nuclear reactions. But the conditions for fusion must be very different from those for standard nuclear reactions.

At least nine nuclear reactions need to be considered as operative in a cold fusion generator. The (p, d) fusion reaction can be represented by the following equation:

$$p + d \rightarrow He3 + E \ (5.5 \ MeV)$$

This is a commonly recognized step in the proton-proton solar fuel cycle. Energy is provided by a 5.5 MeV gamma ray developed by forming helium-3 (He3).

Similarly, d-d fusion can be described by three competing paths shown in the following equations:

$$d + d \rightarrow He3 + n + E \ (3.3 \ MeV)$$
$$d + d \rightarrow T3 + p + E \ (4.0 \ MeV)$$
$$d + d \rightarrow He4 + E \ (23.8 \ MeV)$$

These equations indicate the manner in which protons (p), neutrons (n), tritium (T3), helium-3 (He3), helium-4 (He4), and energy could be produced in fusion reactions. Examples of possible reactions involving these products include:

n + p → d + E (2.22 MeV)
n + d → d + n (elastic)
n + He3 → T3 + p + E (0.764 MeV)
d + T3 → He4 + n + E (17.6 MeV)
γ (> 2.22 MeV) + d → p + n + E (0.5 MeV)

The first reaction, (n, p), depicts neutron capture by protons (e.g., in hydrogen gas or water), producing 2.22 MeV gamma radiation. The second reaction, d(n, d)n, describes neutron scattering from deuterium with a low probability for neutron absorption. Neutron velocity is moderated in the scattering process. The third reaction, He3(n, T3)p, is important for detecting neutrons, since He3 has a high neutron absorption cross section (>1000 barns). The fourth reaction, T3(d, He4)n, is used in some commercial neutron generators and is of more interest to the hot fusion community.

The fifth reaction, d(γ, p)n, indicates that gamma (γ) radiation with an energy greater than 2.22 MeV can cause deuterons to disintegrate. The protons and neutrons produced will each have about 240 keV of kinetic energy. This reaction is one of several sequential steps in a fusion process investigated over the last few years at the National

Aeronautics and Space Administration (NASA) Glenn Research Center.[11] It is of relatively little interest here due to its high level of input energy.

In nuclear reactions, energy can be produced when some of the mass of reactants on the left side of the equations is converted to energy, resulting in products on the right side of the equations. This is a major difference between nuclear reactions and chemical reactions, as little to no mass is converted to energy in chemical reactions. The amount of energy that can be produced by fusion reactions is determined by the difference between the masses using Einstein's formula $\Delta E = \Delta M\ c^2$, where ΔM is the mass difference and c is the speed of light. If mass is given in atomic mass unit (amu), then a mass difference of 1 amu can be converted into 931.5 MeV of energy (1 amu = 1.66054×10^{-27} kg).

Energy and Products from Cold Fusion

The five fusion equations may be grouped together when attempting to understand all possible ways in which energy, helium-3 (He3), helium-4 (He4), protons (p), neutrons (n), and tritium (T3) could be produced in cold fusion:

p + d → He3 + E (5.5 MeV)
d + d → He3 + n + E (3.3 MeV)
d + d → T3 + p + E (4.0 MeV)
d + d → He4 + E (23.8 MeV)
d + T3 → He4 + n + E (17.6 MeV)

The fifth fusion reaction is included due to tritium (T3) from the third reaction and the relatively high amount of energy produced. Probabilities of each reaction in cold fusion are not actually known and could vary with local conditions and amounts of hydrogen and deuterium loading, for example. One can assume that the second, third, and fourth (d, d) reactions occur in somewhat the same manner, and that probabilities of the second and third reactions are about the same due to their physical similarity. Early scientific literature indicates that the probability for observing the fourth reaction should be very low (10^{-7}, or only 1 in 10 million), compared with the second and third reactions.[12, 13] Due to considerations related to conservation of momentum, the alpha particle (i.e., He4) is unable to carry away energy from the reaction as kinetic energy, and the particle cannot be expected to interact with its surroundings. Gamma radiation cannot be observed from excited He4 where the spin and angular momentum are the same as those of He4 in the ground state with zero angular momentum.[14, 15] Some cold fusion scientists, however, believe that this fourth reaction could be most important in producing energy by cold fusion. Energy could be released by several methods discussed below.

It is possible to combine the five reactions, for example, into the following equation to estimate the amount of energy and numbers of particles involved in cold fusion:

8d + p → a [He3 + E (5.5 Mev)] + b [(He3 + n + E (3.3 MeV)] + c [T3 + p + E (4.0 MeV)] + d [He4 + E (23.8 MeV)] + e [He4 + n + E (17.6 MeV)], where a, b, c, d, and e are probabilities of the five different paths and can be adjusted as desired by the interested reader or cold fusion scientists.

Energy, in the form of heat, and helium are two products of cold fusion reactions that have been quantified (and correlated) in experiments. Thus, it is interesting to combine these from the different reaction paths as follows:

$8d + p \rightarrow$ (aHe3 + bHe3 + cT3 + dHe4 +eHe4) + (b + e) n +cp + (5.5a + 3.3b + 4.0c + 23.8d +17.6e) MeV.

Consider a special case where a, b, c, d, and e take on an equal and average value of 20 percent, then a = b = c = d = e = 0.20, and a + b + c +d + e = 1. The energy term would be:

0.20 (5.5 + 3.3 + 4.0 + 23.8 +17.6) MeV.

Also consider a case where tritium from the third reaction above has been converted to He4 by the fifth reaction, that some of the tritium has decayed to He3, and that it is difficult to differentiate between He3 and He4 with the type of mass spectrometer used for the measurement, so that:

$8d + p \rightarrow$ He (combined) + (b + e)n + cp + 10.8 MeV (average).

The reader is encouraged to try other values. For this example, however, about 10.8 MeV of energy would be expected in cold fusion experiments for every helium atom detected. Or, for each watt of excess energy, about 0.58 x 10^{12} helium atoms (per second) should be expected.

Cold fusion scientists have been able since the early 1990s to demonstrate experimentally that a watt of excess power (heat) correlates with production of 0.6 x 10^{12} to 4 x 10^{12} helium atoms per second.[16, 17] Additional amounts of helium may have been produced but not measured if trapped inside the cathode.

Some scientists with suitably sensitive mass spectrometers have also been able to measure helium-3.[18] If only the first reaction (p, d fusion) were considered, approximately 1.1 x 10^{12} He3 atoms would be expected for each watt, but if only the fourth reaction (d, d fusion) were operative, about 0.26 x 10^{12} He4 atoms would be expected for each watt. The amount of helium measured by the mass spectrometer should be smaller than these values, as all helium would not be expected to escape from the cathode.

The following table summarizes the fusion energy calculations:

Examples of Energy from Fusion Reactions

$$M_n = 1.008665 \text{ amu}; \; M_p = 1.007828 \text{ amu};$$
$$M_e = 0.00054858 \text{amu}$$

Reactants(1)	Products(2)	Mass Difference (1-2)	Energy Estimate (mass difference x 931.5 MeV)
p + d → 1.007828 + 2.014102	He3 3.016029	0.005901	5.5 MeV
d + d → 2.014102 + 2.014102	He3 + n 3.01029 + 1.008665	0.00351	3.3 MeV
d + d → 2.014102 + 2.014102	T3 + p 3.016049 + 1.007828	0.004327	4.0 MeV
d + d → 2.014102 + 2.014102	He4 4.00206	0.025602	23.8 MeV
d + T3 → 2.014102 + 3.01605	He4 + n 4.002602 + 1.008665	0.018884	17.6 MeV

By inspecting data in this table, it is possible to conclude that cold fusion reactions are generally able to produce between 3.3 and 23.8 MeV of energy per reaction. Heat could be produced by helium-4, by gamma radiation with He3 (5.5 MeV), and by kinetic energy from helium-3 and neutrons (He3 and n), tritium and protons (T3 and p), and helium-4 and neutrons (He4 and n).

If only the first reaction were considered, note that 5.5 MeV = 8 x 10^{-13} joules. Thus, in comparison with the above discussion on chemical reactions, only 6 x 10^{17} (p, d) reactions would be required to produce 480 kJ of energy. This number of (p, d) cold fusion reactions would be required each second for 480 kW of power.

The (p, d) reaction has been discussed in the paper "A Source of Plane-polarized Gamma-rays of Variable Energy above 5.5 MeV" by D.H. Wilkinson of the Cavendish Laboratory.[19] It indicates that a (p, d) interaction involves a direct radiative transition where the gammas are emitted perpendicular to the path between the proton and deuteron. The paper also indicates that gamma ray energy can be increased from 5.5 MeV by increasing energy of the bombarding protons, and the gammas produced can cause photodisintegration of other deuterons, with the resulting protons emitted along the electric vector.

The photodisintegration effect was first discussed in 1934 by J. Chadwick and M. Goldhaber in "A Nuclear Photo-Effect: Disintegration of the Diplon by Gamma-Rays," and in 1935, they discussed it in greater detail in "The Nuclear Photoelectric Effect."[20, 21] A theoretical discussion was also provided in 1935 by H. Bethe and R. Peierls in "Quantum Theory of the Diplon."[22] Radiothorium producing 2.62 MeV gamma radiation was used, verifying that disintegration can occur when gamma ray energy is greater than the deuteron's binding energy. Energy of separated protons was determined to be 240 keV, and the cross section for photodisintegration was determined to be 0.0001 barn (10^{-28} cm^2). Since mass of the neutron and proton are about the same, their total energy was estimated to be about 500 keV. The deuteron's binding energy was then calculated as (2.62 − 0.50), or 2.1 MeV. By comparison, the deuteron's binding energy from the internet is given as 2.2246 MeV, or 0.002389 amu. A cross section of 0.001 barn (10^{-27} cm^2)

for the reaction is listed in the International Atomic Energy Agency's photonuclear data handbook.[23]

As indicated previously for the fourth reaction, past studies have shown that there is little possibility for gamma radiation to be observed where the spin and angular momentum are the same as those of He4 in the ground state with zero angular momentum. Instead of emitting gamma radiation, however, another way for a nucleus to lose energy and transition from an excited state to its ground state is by internal conversion. This is believed to have been mentioned to Pons and Fleischmann in 1989 by a couple of scientists from the University of Utah.[24] In this case, energy from the excited nucleus is imparted to its own atomic electrons, which are then ejected with high kinetic energies.

Internal conversion has been studied since the 1930s. An early discussion on internal conversion can be found on the internet in "A Theory of the Internal Conversion of Gamma Rays" by H.M. Taylor and N.F. Mott,[25] and an easy-to-read summary is given in "Elements of Nuclear Physics" by Walter E. Meyerhof.[26] Internal conversion is important when the nucleus has a high charge (i.e., high atomic number) and a low (< 1MeV) transition energy is involved, or if (as for excited He4) a radiative transition is inhibited by a special nuclear physics selection rule.

Additionally, electron-positron pairs can be created if the transition energy is high enough.[27] Discussion of internal pair production from 1949 can be found on the internet in

"Life-time for Pair Emission by Spherically Symmetrical Excited State of the O16 Nucleus" by S. Devons et al. and in "Electron Pair Creation by a Spherically Symmetrical Field" by S. Devon and G.R. Lindsey.[28, 29] Internal pair production is important for low atomic number elements and when a high transition energy (>1.024 MeV) is involved. It is particularly important for excited even-even atoms, such as helium-4 and oxygen-16.

This information indicates that kinetic energy of high-speed electrons from internal conversion and pair production (produced by the fourth reaction above) could contribute to heat produced in cold fusion experiments. Some type of electronic compensation may be required for the related electron back current when scaling up to high-power cold fusion generators.

Some cold fusion scientists have devoted significant time and resources to develop other theories besides internal conversion and pair production to explain the reason that gamma radiation is not detected. Such alternative theories are not discussed in this anthology.

Transmutation Reactions

Transmutation is not the focus of this anthology, although the reality of low energy transmutation reactions has been verified by sensitive analytical instruments, such as ion microprobes.[30, 31, 32] The amount of energy per reaction from transmutation is similar to the energy obtained from cold fusion. Operational lifetime can be impacted, however, when cathode material is changed in the process. It is, therefore, of interest that some cold fusion researchers have focused their efforts upon LENR transmutation experiments instead of cold fusion. They seem to be convinced that protons from water in liquid electrolysis experiments and from hydrogen in gas experiments can be made to combine with electrons to form neutral nuclei. The neutralized protons are assumed to enter atoms of cathode material (e.g., perhaps as neutrons), changing the atoms to other elements or isotopes. In the process, material from which cathodes are fabricated is converted (transmuted) to other isotopes or elements. If the cathode were made of nickel, for example, some of the atoms could be transmuted into copper and cobalt. Energy would be produced by the difference between proton and nickel atomic masses and cobalt/copper atomic masses.

Examples of energy from neutron-produced transmutation reactions in the following table can be compared with energies available from cold fusion.

Examples of Energy from Transmutation Reactions

$$M_n = 1.008665 \text{ amu}; M_p = 1.007828 \text{ amu};$$
$$M_e = 0.00054858 \text{ amu}.$$

Reactants(1)	Products(2)	Mass Difference (1-2)	Energy Estimate (mass difference x 931.5 MeV)
n + 58Ni → 59Ni → 1.008665 + 57.935346	e⁺ + 59Co 0.00054858 + 58.933198	0.01026	9.6 MeV
n + 60Ni → 1.008665 + 59.930788	61Ni 60.931058	0.008395	7.8 MeV
n + 61Ni → 1.008665 + 60.931058	62Ni 61.928346	0.011377	10.6 MeV
n + 62Ni → 63Ni → 1.008665 + 61.928346	e⁻ + 63Cu 0.00054858 + 62.939598	-0.003135	None
n + 64Ni → 65Ni → 1.008665 + 63.927968	e⁻ + 65Cu 0.00054858 + 64.927793	0.00829	7.7 MeV

By inspecting data in the table, it is possible to determine that transmutation reactions might, in general, be able to produce from 7 to 10 MeV per reaction. Transmutation reactions that change the cathode's composition, however, should not be relied on for long, consistent generator operation.

3

PHONONS AND THEIR ROLE IN COLD FUSION

Macroscopic View

SCIENTISTS HAVE STUDIED PHONONS (THERMAL VIBRATIONS OF ATOMS IN SOLIDS) since the late 1800s to explain important material characteristics.[33] Phonons are a quantized form of lattice vibration (the energy is carried in discrete quanta). Phonons have energies and vibration frequencies (about 10^{12} to 10^{14} Hertz) related to a material's temperature. Melting can be explained in terms of the increase in number and amplitude of phonons with temperature to the point that the material can no longer remain a solid. Electrical resistance can also be explained in terms of electron scattering by periodic atomic vibrations.

An understanding of thermal vibrations has also been used to describe the manner in which a material's heat capacity varies with temperature up to the temperature reached when all modes of vibration are active, known as the *Debye temperature*. Temperature and heat capacity appear to be important to the cold fusion process. The Debye temperature for nickel sometimes used in cathodes of cold fusion devices is 183 °C (456 °K). For most metals, heat capacity (at medium to high temperature) for a mole of atoms is about 25 joules per degree Kelvin, or 6 calories per degree Kelvin (6 calories/°K).

Some phonon characteristics have been derived from the study of superconductivity.[34] Low temperatures for which materials lack electrical resistance are proportional to Debye temperatures. These transition temperatures were determined in the 1950s to be inversely proportional to the square root of isotopic mass and therefore are related to the atomic vibrations. It was also shown that, at very low temperature, electric current is carried by a pair of electrons. The pair is formed when two electrons are weakly attracted to each other, overcoming their repulsive Coulomb forces as a result of being scattered by phonon vibrations in the material. This is energetically possible because the paired electrons have lost some of their energy and have lower energy than when they were separate. The phonon vibrations, however, have insufficient energy to scatter the more massive pairs, which can then travel through the material almost unimpeded.

The amount of coupling between phonons and electrons in metals is expressed by a coupling constant. The entire phonon spectrum is considered to contribute to the coupling constant, and when the entire phonon spectrum is excited, it is possible to correlate the coupling constant with a metal's kinetic properties. A material's dielectric constant is related to concentration of conduction electrons and to electron-phonon collision frequency. A direct relationship has also been found between the coupling constant and temperature-dependent resistivity.

Microscopic View

Cold fusion scientists have used related concepts since the latter part of 1990 to investigate the idea that cold fusion could be caused by strong interactions between high-frequency phonon vibrations in cathodes of their cold fusion devices and the electrons surrounding deuterium and hydrogen atoms in its channels and defects. The channels and defects are envisioned to be one-dimensional, enabling positive and negative ions to move rapidly toward each other and fuse. When the deuterium and hydrogen atoms lie in these channels or defects, they are considered to have their own spacing in the chain of atoms. The interaction is envisioned to cause some of the electrons to pair up (as indicated above) around the deuterium or hydrogen, producing a negatively-charged atom. The atom with a pair of electrons is more stable than if it had one electron.

Rather than representing electrons in orbits around a nucleus (i.e., the Bohr model), the electric charge of electrons is considered to be in an electron cloud, some of which can pass through the nucleus. A neutron can be produced within an atom in a process called *k-capture* when an electron is captured by a proton in the nucleus.[35] Electron capture can also occur with more distant L-electrons. Electron capture occurs spontaneously, without the addition of outside energy. It is possible for proton-rich isotopes above the line of stability in the chart of the nuclides.[36] For example, a proton in the nucleus of potassium-40 is able to capture an electron, turning it into argon-40. The captured electron is already part of the potassium-40 atom, so the amount of energy produced can be determined just by subtracting the mass of the argon-40 atom (39.962383 amu) from the mass of potassium-40 atom (39.963999 amu) and multiplying the result by 931 MeV/amu. A gamma ray with an energy of about 1.5 MeV is expected. As another example, a proton in the nucleus of beryllium-7 consisting of only four protons and three neutrons is able to capture an electron, turning the beryllium-7 into lithium-7.

The phonon interaction in cold fusion is considered to be strong enough to cause electrons to be absorbed into protons of hydrogen atoms to form neutrons. These neutrons can cause transmutation reactions in adjacent cathode material and would be undesired for long-time operation of cold fusion systems.

By comparison, neutrons can also be produced from nuclear fission or by high-energy particle scattering (such as alpha particles from polonium-210 striking beryllium). The electron in a hydrogen atom normally cannot collapse into the nucleus (it doesn't emit a photon and lose energy) to produce a neutron. And the neutron mass cannot result from combining an electron with a proton, as the rest mass of an electron (0.00054858) in combination with a proton (1.00727647) is insufficient to produce the mass of a neutron (1.008664). An additional 0.00083895 amu or about 780 keV would be required.

Cold Fusion Theory

A theory of cold fusion that appears to be consistent with these ideas from chemistry and physics was detailed in a paper written by Professor K.P. Sinha in June-July 1999, while he was a visiting professor at Harvard University. The theory suggests that cold fusion nuclear reactions can occur as a result of interactions between phonons (high frequency vibrations) in cathode reaction material and electrons (electric charge) of hydrogen and deuterium atoms in defects, cracks and crevices of the reaction material.

A physical description of the process involves the following:

- Hydrogen and deuterium molecules, atoms, and ions are contained by the small defects, cracks, and crevices. When deuterium and hydrogen lie in these channels, they are assumed to be affected by the associated electric potentials within these small volumes and to have their own spacing in the chain.
- The atoms of reaction material can be made to produce optical (high-frequency) phonons.
- The hydrogen species in the channels thermally vibrate with a common frequency as *Einstein*

oscillators. Energies of these vibrations are quantized into levels separated by E = h f, where h is Planck's constant (6.63 x 10⁻³⁴ joule-seconds) and f is frequency in Hertz.

- High-frequency vibrations of cathode reaction material atoms in/near the surfaces of the channels interact strongly by electrostatic fields with electrons of the hydrogen and deuterium, causing the electrons to pair up around individual hydrogen or deuterium atoms. With a pair of electrons, the atom has a negative charge. It is also more stable than if it had one electron.

- An electron or electron pair located on a proton or deuteron and interacting with the phonons can acquire an effective heavy mass, and the corresponding atom or ion is squeezed to much smaller size.

- A resulting negatively charged deuterium or hydrogen ion can strongly attract its complementary positive deuterium ion in a molecule that, for a small instant of time, has no electrons. The electrons can negate the positive coulomb barrier between the ions, enabling the ions to fuse (a comparison can be drawn to muon-catalyzed fusion).

- Since the channels are essentially one-dimension, distantly-spaced positive and negative ions can move rapidly toward each other and fuse.

- A heavy electron or pairs of heavy electrons close to the nucleus can be captured by a proton to form a neutron through an electron capture process. The neutrons can cause transmutation of adjacent reaction material.

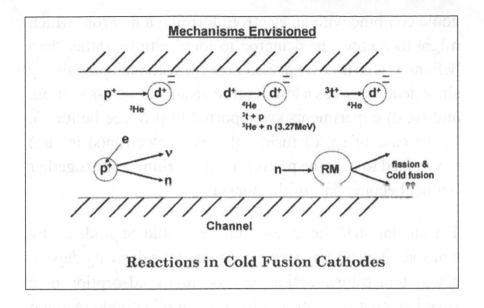

Reactions in Cold Fusion Cathodes

This drawing depicts related reaction mechanisms. The double dash indicates two electrons around a deuteron, causing it to have a negative charge. Examples of *fusion* occur when protons, deuterons, or tritons (tritium) combine with the negatively charged deuteron ions shown on the top row of the drawing.

Fusion of hydrogen (protons, or protium) and deuterium (deuterons) to produce helium-3 (He3) and gamma radiation is emphasized in this anthology since heat produced by absorption of the gamma radiation in the cold fusion system would not be highly localized. Fusion of deuterium (deuterons) and deuterium is second in priority, although the localized, high-energy reaction products could damage the cathode material.

Some researchers have indicated that, instead of the first reaction where (p, d) fusion produces helium-3, the proton could combine with an electron, forming a neutron, which might then enter the deuteron to form tritium, rather than helium-3. This is not expected to occur with high probability since deuterium has a low neutron absorption cross section, and (p, d) experiments are reported to produce helium-3. In the case of (p, d) fusion, the extra electron(s) is (are) envisioned to help the proton and deuterium come together but not become part of the nucleus.

Transmutation in the cathode material could be produced by neutrons. The bottom row, right side of the drawing depicts a way that transmutation may occur by adsorption of a very low-energy neutron into an atom of cathode reaction material (RM) to convert it into another isotope or element. An example is the transmutation of nickel cathode reaction material into cobalt and copper.

The left side of the bottom row depicts electron capture in one of the protons of a hydrogen molecule to produce a neutron (the other proton or atom of the molecule is not shown). The electron is already part of the hydrogen molecule, so the energetics of the process can be estimated by subtracting the mass of the other proton (1.0072647 amu) from the mass of hydrogen (2 x 1.00794 or 2.01588 amu), which takes hydrogen's binding energy taken into account. The difference of 1.0086153 indicates that the hydrogen atom has almost enough mass-energy to form a neutron (1.008664 amu). A mass of 0.000049 amu, however, is still needed, which, when multiplied by 931 MeV/amu, is equivalent to 46 keV. This may be able to be provided by the heavy effective electron mass or phonon energy. Another theory discussed below supports this hypothesis.

Professor Sinha initially suggested the role of electron pairing during a cold fusion meeting held in Bangalore, India in 1989. The idea was also mentioned in an obituary that he wrote for Professor F.C. Frank in 1998. He indicated that he could explain how cold fusion works during a conference hosted by the Integrity Research Institute at the Holiday Inn in Bethesda, Maryland in April 1999. The staff and technical consultants for Epoch Engineering, Inc. in Gaithersburg, Maryland assisted him in further documenting his theory in the summer and fall of 1999. Professor Sinha discussed the theory in a meeting on "The Role of Electron Pairing in Facilitating Fusion, Fission and Other Mechanisms in Reproducible Experiments" held at the Hilton Hotel in Arlington, Virginia on November 18, 1999. About forty copies of his briefing charts were distributed as a technical note to scientists across the nation. Sinha discussed related mathematics in "A Theoretical Model for Low-Energy Nuclear Reactions," published in the *Infinite Energy* magazine.[37] The theory was also discussed in a March 2000 proposal on "New Power Production Technology Reaction Material" to the US Department of Defense.

Professor Sinha continued his theoretical work on cold fusion as a visiting scientist at the Massachusetts Institute of Technology (2000-2003). He met Andrew Meulenberg (PhD, Vanderbilt University in Nuclear Physics) who has been working with him under the aegis of the Science for Humanity Trust in Bangalore, India, which they founded. Since that time, they have co-authored about a dozen related papers and briefings that can be found on the internet. Information on electrostatic fields in the channels was discussed in 2006 and 2007 and additional information on reaction rates was discussed in 2012.

Sinha's theory has several important implications for development and long-time operation of cold fusion systems. Since melting of channels where reactions occur should be prevented, the channels should be designed primarily to support hydrogen and deuterium (p, d) fusion, instead of (d, d) fusion where the resulting nuclear particles could deposit their energy locally. Second, the reaction material should be manufactured to contain a sufficiently large number of extremely small, one-dimensional channels for the power level of interest. The channels should be designed to enable distantly-spaced positive and negative ions to move rapidly toward each other and fuse. Specifications will need to be developed for consistent manufacturing. Another implication of the theory is that cold fusion systems should be made to operate when all phonon modes of vibration are active, that is, above the Debye temperature of reaction material from which cathodes are made. Cathodes should be made of materials that have a high Debye temperature. Also, neutron production and transmutation should be prevented. The relative amounts of gamma radiation, neutrons and transmutation may be able to be regulated by controlling the environment (e.g., temperature and the relative amounts of hydrogen and deuterium in the reaction chamber).

Related theories are discussed in "The Explanation of Low Energy Nuclear Reactions" by Edmund Storms and in "Ultra Low Momentum Neutron Catalyzed Nuclear Reactions on Metallic Hydride Surfaces" by Alan Widom and Lewls Larsen.[38, 39] Widom and Larsen's paper describes how the mass of electrons on reaction material surfaces can be increased by electromagnetic radiation at the surface. The heavy mass electrons can interact with protons and deuterons on the surfaces to produce very low-energy neutrons in a manner similar to the lower left of the above diagram. The low-energy neutrons can be captured by the reaction material, with new isotopes produced that decay by beta emission into other elements.

4

NUCLEAR RADIATION FROM (P, D) FUSION

CHAPTER 2 INDICATED THE FOLLOWING FUSION reactions that are expected to occur in a cold fusion energy generator (although not necessarily with equal probability or the same probabilities as in hot fusion):

p + d → He3 + 5.5 MeV
d + d → He3 + n + 3.3 MeV
d + d → T3 + p + 4.4 MeV
d + d → He4 + 23.8 MeV
d + T3 → He4 + n + 17.6 MeV

The first reaction should produce 5.5 MeV gamma radiation. Some gamma radiation may also be able to be detected as a result of the other reactions but is expected to be of less intensity.

Radiation Measurement Issues

Published cold fusion reports give little confidence that all important factors were considered when measuring high-energy gamma radiation above a few MeVs of energy. Instead of being emitted equally in all directions (i.e., isotropically) from liquid electrolysis experiments, gamma radiation from the first reaction should have a low probability of detection except perpendicular (i.e., $\sin^2\theta$) to a path between interacting protons and deuterons.[40] Computer modeling is needed to determine best placement for the gamma ray detectors.

An even more basic problem, however, is that Geiger-Mueller (GM), sodium iodide, and high-purity germanium (HPGe) scintillator detectors are normally used in laboratory and industrial applications for radiation decay energies that are less than a few MeV. Detection of higher-energy gamma radiation is rarely mentioned in cold fusion literature.

Some cold fusion experiments were interested in detecting lower-energy gamma radiation produced by neutron activation. The work on cold fusion sponsored by the Electric Power Research Institute (EPRI), for example, used a thin-window, Compton-suppressed, HPGe gamma-ray spectrometer to ensure low background for x-rays and gamma radiation below a few MeV. It had a detection efficiency of only 1 percent at 1.3 MeV, and even lower detection efficiency for higher energies. EPRI's report indicates that constraints and demands of gamma counting prevented a successful outcome.[41] Upgrades using higher atomic number detector materials, such as bismuth germanate oxide (BGO) and lanthanum bromide (LaBr3) could have been used for higher energies but are very expensive.

Utility of (p, d) Fusion

For every watt of heat energy, 1.1×10^{12} gamma ray photons per second should be expected if the heat were caused entirely by gamma radiation from the first reaction. About 2×10^{17} reactions per second are needed for a generator that produces 200 kW. Radiation would not be emitted when the generator is not operating, such as during maintenance.

One of the reasons for focusing on (p, d) reactions is that they should occur more easily in a cold fusion environment than deuteron-deuteron (d, d) reactions. This is discussed in "Radiative Proton-Capture Nuclear Processes in Metallic Hydrogen," written by Setauo Ichimaru and published in *Physics of Plasmas*.[42] A second reason, mentioned earlier, is that facilitating (p, d) reactions, for example, over (d, d) reactions, should help to reduce the amount of local heating in each of the very small, microscopic-sized reaction sites within the cold fusion generator cathode. Deuteron-deuteron reactions produce neutrons, protons, and other nuclear particles that have kinetic energies high enough to damage the cathode material. Gamma radiation from (p, d) fusion would be distributed throughout a cold fusion generator's mass and would not be absorbed significantly within these local reaction sites. As a result, the cold fusion generator should be able to operate for much longer periods before maintenance is required on the cathode.

Gamma Ray Attenuation

In the process of producing heat, gamma radiation from (p, d) reactions can be expected to ionize thousands of atoms and molecules in the cathode and other parts of the generator. Only 10 to 1000 eV are needed for each ionization.

The relation between energy and frequency of the radiation is described by the formula $E = h\,f$, where E is energy in joules, h is Plank's constant (6.63×10^{-34} joule-seconds), and f is frequency of the radiation in Hertz. Thus, 5.5 MeV gamma radiation photons will have a frequency of 1.3×10^{21} Hertz. Note that this is speaking about *photons* or electromagnetic radiation, not *phonons*. Wavelength (λ) of the radiation can be determined from the formula $\lambda = c/f$, where c is a constant equal to the speed of light (3×10^{8} meters per second). The 5.5 MeV gamma ray will have a wavelength of 2.3×10^{-3} Angstroms. This indicates a low probability of interaction between the radiation and any single atom or molecule in the cathode and other parts of the generator. The generator is composed of a great many atoms of material with which to interact. Higher-energy gamma ray photons would have less probability of interacting with the material.

Gamma radiation can be attenuated by the photoelectric effect (most important for gamma energy below several hundred keV), by Compton scattering (most important for gamma energy between several hundred keV and a few MeV), and by pair production (considered for gamma energies above 1.022 MeV). Each of these effects will come into play for 5.5 MeV gamma ray photons produced by (p, d) fusion reactions. Each involves scattering of electrons in generator construction materials, and heat is produced as the electrons lose their energy by Coulomb interactions with atoms in the material.

In the photoelectric effect, a gamma ray photon interacts with an atom of absorbing material and results in ejection of an electron from the atom. The ejected electron receives all of the energy of the gamma ray minus the electron's binding energy in the atom, and its scattering may induce secondary ionization events in the material. The probability of the photoelectric effect is proportional to atomic number (Z) of the absorbing material and is inversely related to gamma ray energy. The photoelectric effect is most important for low-energy gamma rays interacting with heavy elements.

Compton scattering also involves the interaction of gamma ray photons with atoms of a material and ejection of electrons from the interaction. In Compton scattering, however, only a portion of the energy from the higher-energy gamma ray is transferred to the electron, and the remaining energy is transmitted as gamma rays at lower energy. As with the photoelectric effect, the probability of Compton scattering is proportional to atomic number (Z) of the absorbing material and is inversely related to gamma ray energy. Compton scattering produces a continuum of scattered gamma ray energies from 250 keV below the highest energy of the incident gamma radiation (known as the "Compton gap") down to a minimum value. The minimum energy (in keV) of gammas produced by Compton scattering can be determined from the following equation:

$$E_{min} = 511 \, E_{incident} / (511 + 2 \, E_{incident}).$$

In pair production, a gamma ray photon above 1.022 MeV can be converted into an electron-positron pair near the nucleus of an atom of the absorbing material. Any energy of the incident gamma ray photon greater than 1.022 MeV is transferred to the electron and positron as kinetic energy. The electron and positron can produce additional ionization in the absorber material. The positron will eventually be annihilated, producing two 511 keV gamma rays, which will interact further with the material.

Additional information on these effects can be found, for example, in a textbook by Glenn Knoll, *Radiation Detection and Measurement*.[43] The National Institutes of Standards and Technology (NIST)'s XCOM database may be used in determining the amount that gamma radiation is absorbed in various materials.[44] In addition, the specific amount of absorption by the photoelectric effect, Compton scattering, and pair production can be determined from any of several x-ray and gamma ray calculators on the internet. These data can be used to determine the required wall thickness for cold fusion generator components.

Radiation Safety

Radiation safety is an important consideration for almost any cold fusion generator power level. Radiation dose can be estimated from the definitions of "gray" (Gy) and "rad" (radiation absorbed dose):

$$1 \text{ Gy} = 100 \text{ rad} = 10^4 \text{ ergs/gram} = 1 \text{ Joule/kg} = 6.24 \times 10^{12} \text{ MeV/kg}$$

If one watt of heat and 1.1×10^{12} gamma rays per second were produced by the cold fusion generator, then one meter away without shielding, there would be a flux of 10^7 gamma rays/per second through each square centimeter of area.

Radiation dose rate can be estimated as follows, assuming that human tissue has a density of about 1.1 gram/cm^3 and about 8 percent of the gamma radiation will be absorbed by 2.0 cm tissue thickness (i.e., 92 percent transmitted through the tissue):

$$[0.08 \times 10^7 \text{ photons/second} \times 5.5 \text{ MeV/photon}] / [1.1 \text{ gm/cm}^3 \times 2.0 \text{ cm} \times 1 \text{ cm}^2]$$

$$= 2 \times 10^6 \text{ MeV/sec/gm} = 2 \times 10^9 \text{ MeV/sec/kg}$$

Dose rate a meter distance from the radiation source is therefore estimated as 0.3 mGy/second or 0.03 rad/

sec. This is high enough for concern, as it is about three times the radiation received from a chest x-ray. In three hours of exposure (about 10,000 seconds), this would amount to 300 rad and could result in serious health consequences. For their safety, researchers should not be permitted to stand close to active experiments. Experiments producing less than a watt would produce comparatively less radiation, and the amount of radiation received can be significantly reduced with shielding and distance.

Radiation Safety in 200 kW Generator Design

A cold fusion generator can be designed to absorb most of the nuclear radiation that it produces. Approximately 2×10^{17} (p, d) fusion reactions per second are needed in the cold fusion generator for 200 kW of power. If the walls of the reaction chamber and heat exchanger are sufficiently thick to attenuate about 90 percent of the radiation and the cathode were encased in a tungsten sleeve to increase this to 98 percent, then 2 percent of the photons (4×10^{15}) would still be emitted into the area around the generator. One meter away, without shielding, a flux of 10^{11} gamma rays per second could be expected through each square centimeter of area.

If 8 percent of the gamma radiation were absorbed by 2.0 cm thickness of human tissue (i.e., 92 percent transmitted through the tissue), the radiation dose rate is estimated as:

[0.08 x 10^{11} photons/second x 5.5 MeV/photon] / [1.1 gm/cm^3 x 2.0 cm x 1 cm^2]

= 2 x 10^{10} MeV/sec/gm = 2 x 10^{13} MeV/sec/kg.

Dose rate one meter from an operating 200 kW generator would be about three (3) Gy per second (300 rad/sec). The generator would need to be housed in a protective compartment to provide extra shielding and a safe keep-out distance for operator safety.

Nuclear radiation is a hazard because its energy (e.g., MeVs) is large enough to produce ionization in living tissue. A linear relationship exists between radiation dose and probability of damage to the human body, and there is no safe threshold of radiation. Radiation from a cold fusion system, therefore, must follow the principle of *as low as reasonably achievable* (ALARA). The generator should be designed so that gamma radiation produced in the generator's cathode during operation will be absorbed by its reaction chamber and heat exchanger walls. Any radioactive gas (tritium) should be prevented from leaking into the environment by a header that can be completely sealed to the reaction chamber. Operation and maintenance procedures should ensure that radioactive gas is removed from the reaction chamber before the header is opened, and that cathode removal and replacement during maintenance are performed in a manner that prevents exposure to personnel.

Important safety advantages can be provided by cold fusion generators compared with other nuclear power systems.

Intense gamma radiation would not be emitted when the generators are powered down for maintenance or in a standby mode. For space power applications, astronauts would be able to perform maintenance and other activities in the same area of their spacecraft where the cold fusion generators are located. By comparison, although a NERVA-type fission reactor burning several hours might provide the necessary velocity for a trip to Mars, it would emit intense radiation not only when operating but also in standby mode after being operated once. The spent nuclear waste would emit harmful levels of radiation for decades or centuries.[45] Astronauts would be confined behind a heavy radiation shield *at all times* and not able to perform required functions without receiving a grave radiation dose. A fission-type reactor presents other serious operational problems. When not operating before launch, a fission reactor will need to employ sure safe controls to prevent it from becoming critical. Fission space reactors can also emit highly radioactive particles into the atmosphere and space along with propellant. All surfaces that the particles contact would become contaminated, as well as the earth's atmosphere.

The number of photons absorbed by the walls of a cold fusion reactor and its heat exchanger is proportional to the number of photons produced in the cathode (I_0) and to the thickness of the walls. The transmitted intensity (e.g., number of photons) not absorbed in traveling a distance (x) through its walls can be determined from the formula:

$$I(x) = I_0 \exp(-\mu x),$$

where μ is the linear attenuation coefficient of radiation of a particular energy through the absorbing material. This expression indicates that the intensity of transmitted radiation will decrease in an exponential fashion with the thickness of the absorber and with the rate of decrease controlled by the linear attenuation coefficient. The intensity decreases from I_o, the number of photons at $x = 0$, in a rapid fashion initially and then more slowly in an exponential manner. The units of μ are the reciprocal of those used for x. Values of μ are in units of (cm^{-1}) when x is in (cm). The linear attenuation coefficient is characteristic of individual absorbing materials. It is used when different thicknesses for an absorbing material of the same density are being considered, and increases as the atomic number of the absorber increases. Low atomic number (Z) materials (e.g., carbon) have a small value and are easily penetrated by gamma rays. Heavier materials (e.g., lead) have a large linear attenuation coefficient and are relatively good absorbers of radiation. The linear attenuation coefficient also decreases as gamma ray energy increases. The linear attenuation coefficient is 0.24 cm^{-1} for 5.5 MeV gammas through stainless steel (iron). An additional energy shield made of a high atomic number material such as tungsten can be used around the cathode. The linear attenuation coefficient is 0.80 cm^{-1} for 5.5 MeV gamma rays through tungsten.

The linear attenuation coefficient divided by density (ρ) of the absorber material is known as the *mass attenuation coefficient*. If the linear attenuation coefficient is in units of (cm^{-1}) and density of the material is in (gm/cm^3), then the equivalent mass attenuation coefficients is in units of (cm^2/gm). The mass attenuation coefficient is 0.0309 $cm^2/gram$ for 5.5 MeV gammas through stainless steel and 0.0415 cm^2/gm for 5.5 MeV gammas through tungsten. Mass attenuation coefficients can be found, for example, in tables from the National Institute of Standards and Technology.[46]

Radiation Exposure

Exposure from gamma and x-ray radiation emitted into the air is measured by a unit called the roentgen (R). This is viewed as the quantity of energy deposited, as determined by the amount of ionization produced in the air by the radiation. It applies only to gammas and x-rays and not to other types of ionizing radiation. It is not generally used to describe any effect that would be produced in biological systems since it refers to the effect in air from radiation passing through it. Radiation exposure in roentgens is measured with a survey meter (or "Radiac").

The roentgen is defined as the quantity of radiation that liberates one electrostatic unit (esu) of electricity by ionization in one cubic centimeter of dry air at 0°C and one atmosphere of pressure (density = 1.29 gms/liter). The electrostatic unit of charge is equivalent to releasing 3.33×10^{-10} Coulomb of electricity per cm^3, where one Coulomb contains 6.25×10^{18} charges. A roentgen is also considered as the amount of radiation that produces either 2.08×10^9 ion pairs per cm^3 or 2.58×10^{-4} Coulomb in one kilogram of air under these conditions. One roentgen produces about 88 ergs of energy in one gram of dry air at standard temperature and pressure (STP). This is equivalent to the absorption of 0.00877 gray (0.877 rad) in dry air. Roughly, therefore, 1 roentgen of exposure could result in the absorption of almost 1 rad in the human body.

Absorbed Dose

The effect of radiation on a biological system depends upon the amount of radiation energy that has been absorbed. Radiation dose is the amount of energy deposited into a given mass by ionizing radiation. As indicated above, units for dose are the gray (Gy) and radiation absorbed dose (rad):

$$1 \text{ Gy} = 100 \text{ rad} = 10^4 \text{ ergs/gram} = 1 \text{ Joule/kg} = 6.24 \times 10^{12} \text{ MeV/kg}$$

These units of absorbed dose apply to all types of ionizing radiation, to include alpha and beta particles and neutrons as well as to gamma radiation and x-rays.

Equivalent Dose

The amount of absorbed dose that produces a particular effect in biological systems can vary appreciably from one type of radiation to another, as well as with energy. The "rem," which stands for "roentgen equivalent for man," is a unit of measurement that relates the absorbed dose to its biological effect. The international unit (SI) is called a "sievert" (Sv) (1Sv = 100 rem = 1 Joule/kg), similar to the use of Gy and rad (1Gy = 100 rad = 1 Joule/kg). The rem and sievert are units for "equivalent dose" because they are determined on the basis of weighting factors relative to types and energies of the radiations. The weighting factor was previously called a "quality factor" and is a ratio of gamma (or x-ray) dose that would be required to produce the same biological effect as the other type of radiation in question. "Relative biological effectiveness" (RBE) is a term used by the military instead of "weighting factor." Gammas and x-rays produce the least damage and are assigned a weighting factor of one, whereas alpha particles and fission products produce the most biological damage and are assigned a weighting factor of twenty. The equivalent dose in sieverts is calculated by multiplying absorbed dose in grays by the weighting factor:

Equivalent dose in biological tissue (sieverts) = absorbed dose (grays) x weighting factor.

The roentgen, therefore, is a measurement of radiation exposure in air; the gray or rad is a measurement of the radiation absorbed by a material or tissue; and the sievert or rem is a measurement of the biological effect of the absorbed radiation. In general, values for the roentgen, rad, and rem may be considered to be approximately equivalent.

Medical personnel are trained to limit radiation exposure and ensure that health benefits of radiation outweigh the risk of exposure. A dose of 100 mSv presents a cancer risk. A typical person's equivalent dose is about 2–3 mSv per year from all sources of radiation, including cosmic radiation, medical x-rays, and natural radioactivity. The amount of radiation from medical procedures is estimated as: 10 mSv from a full-body CT scan; 0.4 mSv from a mammogram; 0.1 mSv from a chest x-ray and 0.01 mSv from a dental x-ray. International standards recommend that annual exposure for workers around radiation sources be limited to a whole-body dose less than 20–50 mSv/yr (2–5 rem/yr). A single dose of 1000 mSv (100 rem) will cause radiation sickness and nausea. By comparison, the dose rate for gamma radiation one meter from a 200 kW (p, d) cold fusion generator is estimated to be about three sieverts per second (300 rem/sec). While the generator is operating, astronauts would need to be protected by a radiation shield and sufficient stand-off distance.

Comparison to Other Radiation Sources

Several other sources of radiation need to be considered. First, although (d, d) fusion is less probable in a cold fusion environment than (p, d) fusion, the (d, d) reactions that occur are expected to produce neutrons and tritium. Tritium decays with a half-life of 12.3 years to helium-3 by emitting beta electrons, which are a hazard for the skin, eyes, and internal organs. If one-tenth of cold fusion reactions in the cathode involve (d, d) fusion and half of these produce tritium, then 1×10^{16} atoms or 5×10^{15} molecules of radioactive tritium would be produced per second. Over a year of operation between maintenance periods, about 0.3 moles of radioactive tritium gas would need to be recovered safely. Also, induced radioactive products can be produced in the cathode. Appropriate cathode handling procedures will be needed to prevent exposure of personnel and contamination of the environment.

It is possible to compare radiation from induced radioactive products in the cathode with dose from industrial and medical sources. For example, cesium-137 with a half-life of 30.2 years emits 0.662 MeV gamma rays. Cobalt-60 has a half-life of 5.27 years, and decaying to nickel-60, emits 2.5 MeV (total from 1.33 and 1.17 MeV gamma rays). Iridium-192 with a half-life of seventy-four days emits 0.31, 0.47, and 0.60 MeV gamma rays. Intensity (or "activity") of these sources is determined by the decay equation:

$$dN/dt = -\lambda N,$$

where the decay constant, $\lambda = 0.693/(\text{half-life})$. The intensity/ activity of a source (dN/dt) is expressed as the number of disintegrations per second. The curie (Ci) is sometimes used as the unit of activity for radioactive source materials and is defined as the quantity of radioactive material in which 3.7×10^{10} atoms disintegrate per second. This is approximately the amount of radioactivity emitted by one gram (1 g) of radium-226. The becquerel (Bq) is a small unit for activity in the international system (SI) and is defined as the quantity of radioactive material in which one atom is transformed per second. Since it is a small unit of measurement, kilobecquerels (kBq), megabecquerels (MBq) and terabecquerels (TBq) are normally used. One curie equals 3.7 x 10^4 MBq. Specific activity of Cs-137 is 3.215 TBq per gram (87 Ci/gm). Cobalt-60 has a specific activity of 44 TBq per gram (1100 Ci/gm), and Ir-192 has a specific activity of 341 TBq per gram (9.22 KCi/gm).

Comparison to NERVA and Cobalt-60 Heat Sources

Cold fusion generators may provide an attractive option to power deep space and planetary missions. In comparison, for example, to NERVA-type reactors and cobalt-60 heat sources, intense radiation would only be produced during periods when the cold fusion generators are powered up.

Cobalt-60 is particularly important in medical and industrial applications due to its relatively long half-life compared to other gamma ray sources. It is used in radiotherapy cancer treatment, food sterilization, and in nondestructive detection of structural flaws in metal parts. The US Department of Energy during the 1960s and 70s investigated the possibility of using cobalt-60 heat sources for remote power in space.[47, 48]

Since the density of cobalt is about 9 gms per cm^3, 1 curie = 3.7 x 10^{10} disintegrations/sec, 1 MeV = 1.6 x 10^{-13} joule, and one joule/sec = 1 watt, the power density from a cobalt-60 heat source with an activity of 350 Ci/gm (instead of 1100 Ci/gm indicated above) can be estimated by the following calculation:

350 Ci/gm x 9 gm/cm^3 x 3.7 x 10^{10}/sec-Ci x 2.5 MeV (avg) x 1.6 x10^{-13} J/MeV

= 47 watts/cm^3 (thermal)

Internal radioactive components for a 30 kW cobalt-60 heat source (5.8 kg) would occupy about 640 cm^3. An imaginary sphere with a meter radius surrounding this source would have a volume of 4 x 10^6 cm^3 and a surface area of 1.3 x 10^4 cm^2. The exposure rate one meter from an unshielded source can be estimated as:

350 Ci/gm x 5800 gms x 3.7 x 10^{10}/sec-Ci x 2.5 MeV (avg)]/ [1.3 x 10^4 cm^2]

= 1.4 x 10^{13} MeV/sec/cm^2

As above, human tissue can be assumed to have a density of about 1.1 gram/cm^3, and 8 percent of the gamma radiation can be assumed to be absorbed by 2.0 cm of the tissue. Absorbed tissue dose from the unshielded source is estimated as:

[0.08 x (1.4 x 10^{13} MeV/sec)]/ [1.1 gm/cm^3 x 2.0 cm x 1 cm^2]

= 5 x 10^{11} MeV/sec/gm = 5 x 10^{14} MeV/sec/kg, or about 10,000 rad per second.

One second of exposure from an unshielded 30 kW cobalt-60 source would be lethal.

A heavy enclosure would be necessary to protect nearby personnel. The above references indicate that a 1.75 cm thick, spherical tungsten shield (density 19.3 gms/cm^3) would capture about 90 percent of the radiation (10 percent escape) and would reduce the absorbed dose rate in air ten feet away to about 2 rad/sec. An operator at this distance would receive a dose of 600 rad in five minutes. A nearly solid, 18.7-inch diameter spherical tungsten enclosure (weighing over 2,200 pounds) could be used to reduce the dose rate to 0.1 rad per hour at a distance of one meter.

The NERVA nuclear space engine discussed earlier was developed under the Rover/NERVA program between 1955 and 1972. It contained uranium-graphite composite reactor fuel and would require a very heavy, 10,000 pound shield between the reactor and flight cockpit to protect the astronauts.[49] When operating at full power, the dose rate from gamma radiation at the surface of the reactor (ten feet from the center of the reactor core and at its midplane) would be 30,000 rad per second. Astronauts flying (e.g., in another spacecraft) as close as ten miles to the side of NERVA operating at full power would receive a radiation dose of 4 rem per hour. After a day in coast or shutdown mode following thirty minutes at full power, gamma radiation dose rate at the surface of the reactor (ten feet from the center of the reactor core at its midplane) would be 4.7 rad/sec. A couple minutes of exposure for an astronaut approaching the side of the reactor would be fatal.

5

SCALE-UP AND INDUSTRIALIZATION

NDUSTRIALIZATION OF COLD FUSION IS EXPECTED TO BE
a difficult undertaking due to the complexity of
subsystems to be built, integrated, and tested. Significant
urgency is needed due to the climate crisis. This technology,
however, does not appear to be as technically complex as
past development of nuclear fission or past and present
hot fusion work. A Manhattan Project approach is not
expected to be required in terms of development expertise
and complexity.

The development process must focus upon new energy
concepts that are highly different from those for nuclear fission,
and many areas of expertise are involved. It requires scientists
and engineers with advanced knowledge and understanding
of physics and engineering and who are highly experienced,
team oriented, and committed to further innovation. Rapid
prototyping can help to shorten development timelines.
Industrial partners will be technically advanced research and

development companies, highly interested in solving the climate crisis, and committed to advancing scientific discovery and technical innovation. Many advanced development companies and institutions today, by comparison, are specialized and limited in the required areas of expertise. Some type of joint development program, therefore, will be required to integrate work of the team members.

The program should begin from the present compendium of cold fusion knowledge and proceed rapidly through well-organized stages of advanced development and demonstration, followed by production and deployment. It will need to incorporate modern information-management features. Financial and technical support from various entities interested in its success should be rapidly integrated, including those beyond the United States, as approved by appropriate federal agencies. Due to its critical nature, participants must strongly support a peer review/oversight process conducted by other experts from the scientific community and be *transparent* to them in revealing program activities and technical progress. This requirement has generally not been a priority of cold fusion scientists to date. Agreements regarding laboratory quality assurance will be needed to ensure information accuracy and consistency. The program will also need to be operated independently from, and noncompetitively with, other energy development programs, to include hot fusion's influence.

A continued commitment to further innovation is essential to facilitate integration of additional valuable technical information discovered as the program proceeds. Expansive business opportunities can be created for the requisite technically advanced R&D companies that participate in this crucial initiative. New components and component improvements can be expected, such as more robust, higher-efficiency cathodes and materials for cathode construction. Advanced modeling should result in improved understanding of phonon processes in the cathode, energy produced in individual reaction material sites, and heat flow through the reaction vessel and into a boiler/heat exchanger. New methods are also needed for data analysis to quantify reaction products in the cathode. Sophisticated electronic subsystems will be needed for remote control of generator operation. Many such possibilities will lead to intellectual property development. Confidentiality of new inventions will be protected through nondisclosure agreements (NDAs) and the patent application process.

Problems with Hot Fusion

Work has been supported by the federal government on hot plasma fusion (i.e., "hot fusion") in government laboratories, academia, and private industry since the mid-1950s. Hot fusion, however, is not expected to become practical for many more decades. The US Department of Energy initiated Project Sherwood (1957) to coordinate the work and demonstrate that nuclear fusion reactions could be made to occur in high-temperature deuterium and tritium gas plasma and that the heat produced could be used to generate electricity. Hot fusion development, however, has a number of drawbacks:[50]

- The machines are expensive to build and operate.
- Their plasmas are unstable and turbulent.
- The plasma is heated with waves propagating through inhomogeneous, active media.
- Performance is strongly affected by plasma interactions with internal material surfaces.
- The shape of the magnetic field is critical to operation but determined by competing factors.
- The plasma erodes the internal wall, and plasma becomes contaminated.
- The internal wall suffers radiation damage.
- Tritium used in the reactors is radioactive, presenting a serious health hazard.
- Thermonuclear energy has to be confined for seconds to maintain nuclear burn.

Hot fusion requires gas ion temperatures greater than 100 million degrees. The hot plasma must satisfy the Lawson criteria (or "triple product") obtained when multiplying temperature, confinement time and plasma density together.[51] When deuterium and tritium are used, the Lawson criteria is required to be at least 3×10^{21}, in units of KeV-s-ions/m^3, or 3.5×10^{28}, in units of °K-s-ions/m^3. Early hot fusion machine designs (e.g., the stellarator, toroidal pinch, and magnetic mirror) attempted to solve plasma instability problems preventing required temperatures from being achieved. Other reactor designs, such as the tokamak ("toroidal chamber with magnetic coils"), were also invented to solve the instability issues, and the required ion plasma temperature was reached in the mid-1990s. Other problems with hot fusion are yet to be completely solved. These are discussed, for example, in *Fusion Science and Technology* published by the American Nuclear Society (ANS).[52]

The hot fusion community is relying on next development steps with a large experimental tokamak, called the International Thermonuclear Experimental Reactor (ITER) being built in Cadarache, France.[53] The ITER has a plasma volume of 840 m^3 and is designed to produce 500 megawatts (MW) or 590 kW/m^3 of thermal output power but only for about twenty minutes. Since 1 MeV = 1.6 x 10^{-13} joule, if each deuterium-tritium (d, t) interaction were to fuse and produce 17.6 MeV of energy, the density of deuterium and tritium would need to be about 2 x 10^{17} ions/m^3. Most deuterium-deuterium interactions are not expected to fuse, and helium produced will affect reaction efficiency. Additional information about the ITER can be found in "A Star in a Bottle," published in the *New Yorker*.[54] Breakeven, with fusion power exceeding the power required to heat and sustain the plasma, is estimated to take several more decades and cost billions of dollars. If deuterium were solely used, instead of radioactive tritium with deuterium, the reactor would need to operate at a much higher (and presently unattainable) plasma temperature.

Compact fusion reactors have also been designed with significantly smaller plasma volumes than the ITER tokamak. Lockheed Martin, for example, recently announced that it is developing a compact hot fusion reactor that combines promising aspects from several earlier designs.[55] While the smaller machines are easier to build and test, and less expensive than the larger ITER tokamak, *none have yet reached breakeven.*

Problems with Fission Power Plants

Recent state and national newspapers have indicated that the nuclear power (fission) industry is in trouble. Several construction projects are bankrupt. Nuclear industry representatives apparently have withheld information over the years on the need to solve plant safety issues and off-site radiation released into local communities. According to Gregory Jaczko, chairman of the Nuclear Regulatory Commission (2009–2012), "(Nuclear fission) is no longer a viable strategy for dealing with climate change, nor is it a competitive source of power. It is hazardous, expensive and unreliable, and abandoning it wouldn't bring on climate doom. The real choice now is between saving the planet and saving the dying nuclear industry."[56]

US Government Participation

Agencies of federal and state governments, especially large power users and those with an R&D mission for energy development, should support cold fusion systems development. Public-private partnerships are needed between industry and these agencies. The US Navy has supported research in this area of technology, and more recently, some support has been provided by the National Aeronautics and Space Administration.

The US Department of Energy, previously known as the Atomic Energy Commission, is expected to be a key participant, as it was established to direct research and development on peaceful uses of nuclear energy (as well as to control development of nuclear weapons). The Atomic Energy Act (1946) established a policy for developing and using atomic energy, improving the public welfare, increasing the standard of living, strengthening free competition in private enterprise, and promoting world peace. The act provides for programs that assist and foster private R&D, encourage maximum scientific progress, and control and share information concerning the practical, industrial applications of atomic energy. It provides for federal R&D support to ensure adequate scientific and technical progress, plus a system of administration that enables Congress to be informed so that it can support required legislative action. The Nuclear Regulatory Commission (NRC) will also need to be involved from a nuclear health and safety standpoint.

Supporting Patent Knowledge Base

Scientific understanding of cold fusion has been discussed by scientists worldwide over the last thirty years, and intellectual property regarding cold fusion and LENR has been protected in the US since 1994 through the patent approval process. Much of this information is available in technical reports on the internet and from the US Patent Office's website. These sources can be used to support a cold fusion industrialization program. The following are examples:

a. 5,318,675, "Method for Electrolysis of Water to Form Metal Hydrides," June 7, 1994, by James A. Patterson: describes a type of liquid electrolysis device containing microspheres coated with a conductive palladium layer and an electrolyte composed of water or heavy water and a conductive salt (e.g., lithium sulfate). Methods are discussed for assembling the electrolysis device, electrolyte and microspheres. Test setup and results are also discussed. Patent examiner: Donald R. Valentine. Application was dated July 20, 1993.

b. 5,411,654, "Method of Maximizing Anharmonic Oscillations in Deuterated Alloys," May 2, 1995, by Brian Ahern et al.: developed with U.S. Air Force support; describes concepts for a liquid electrolysis device containing either deuterium or hydrogen (sublattice) in many small regions on surfaces of a palladium-silver, palladium or nickel cathode (host lattice), at a ratio of at least 5 atoms of the sublattice to 10 atoms in the host lattice and energized by low frequency (5-2000 Hz) voltage. A theoretical discussion is provided on enhancing deuterium or hydrogen oscillations within the small cathode regions. The deuterium or hydrogen is provided to the cathode by electrolysis of water or heavy water. Methods of making many small regions on host lattice surfaces (e.g., scribing and layering) are also discussed. Patent examiner: Donald R. Valentine. Application was dated July 2, 1993.

c. 6,248,221 B1, "Electrolysis Apparatus and Electrodes and Electrode Material Therefor," June 19, 2001, by Randolph R. Davis et al.: discusses design of a gas or gaseous type of cold fusion device with cathode reaction material comprised of nanocrystalline (e.g., nickel) particles, a porous ceramic reaction vessel between the anode and cathode, a microwave waveguide starter/initiator, and a relatively simple electronic control circuit. A theoretical discussion is provided on the movement of hydrogen into and through the cathode by electrolysis, gas pressure, and electric and thermal diffusion. Spray conversion processing is discussed as a method of making nanocrystalline particles for the cathode. Patent examiners: Kathryn Gorgos and Thomas H. Parsons in USPTO Art Unit P/1729. Application was dated June 1, 1999.

d. 7,244,887 B2, "Electrical Cells, Components and Methods," July 17, 2007, by George H. Miley: describes concepts for a wet (or dry) electrolysis type of cold fusion device that uses a multi-layer thin film cathode made of palladium, titanium or nickel, for example, or metallic nanoparticles. Cell designs which employ loading of ionic hydrogen from a hydride storage layer are considered. Results from experiments with multi-layer thin films are also included. Patent examiner: Bruce F. Bell. Application (under the international Patent Cooperation Treaty or PCT) was dated February 26, 2001.

e. 7,893,414 B2, "Apparatus and Method for Absorption of Incident Gamma Radiation and Its Conversion to Outgoing Radiation at Less Penetrating, Lower Energies and Frequencies," February 22, 2011, by Lewis G. Larsen and Allan Widom: describes concepts for a gas type of device to produce heavy electrons within oscillating "surface plasma polaritons (SPPs)" on metal substrate (e.g., nickel) surfaces that can interact by extremely intense electric fields directly with oscillating protons (or deuterons without protons) and be captured by the protons (or deuterons) to form low energy neutrons. The neutrons can be captured by device construction material (e.g., palladium-lithium alloy) in a low energy nuclear reaction (LENR) process, transmuting the material and producing energy. A theoretical discussion is provided on the manner in which gamma radiation from the LENR reactions, or from outside sources, may be shielded by SPP electrons that absorb gamma ray electromagnetic energy. Methods of making the metallic working surface are also discussed. Patent examiners:

Robert Kim and Hanway Chang in Art Unit P/2881. Application was dated September 9, 2005.

f. 8,227,020 B1, "Dislocation Site Formation Techniques," July 24, 2012, by George Miley: describes concepts for a gas type of cold fusion device to use multi-layer thin films made of palladium, titanium or nickel, for example, or metallic nanoparticles and providing dislocations where reactions would occur. Theoretical discussion on pycnonuclear reactions and reaction rate equations are provided. Detailed discussion is provided on results from experiments with multi-layer thin films. Methods are provided to increase the density of dislocations in the thin films. Patent examiner: Brian K. Talbot in Art Unit P/1715. Application was dated March 31, 2008.

g. 8,440,165 B2, "Dislocation Site Density Techniques," May 14, 2013, by George Miley and Xiaoling Yang. Information in this patent is similar to that in 8,227,020 B1. Patent examiner: Frank Lawrence, Jr. in Art Unit P/1776. Application was dated March 7, 2012.

h. 8,419,919 B1, "System and Method for Generating Particles," April 16, 2013, by Pamela A. Boss et al.: developed with US Navy Department support; describes design of a liquid electrolysis device whose cathode is formed by co-deposition of deuterium and a deuterium absorbing metal, such as palladium. Steps in making and operating the device are discussed, to include the use of CR-39 plastic as a detector. Results are provided from related experiments. Patent examiners: Keith Hendricks in Art Unit P/1773 and Steven A. Friday in Art Unit P/1795. Application was dated September 21, 2007.

i. 8,603,405 B2, "Power Units Based on Dislocation Site Techniques," December 10, 2013, by George Miley and Xiaoling Yang. Information in this patent is a continuation-in-part and similar to 8,227,020 B1 and 8,440,165, but with additional system design information provided for small power units (Figure 16) and gas-loaded reaction generator modules (Figure 17). Patent examiner: Frank Lawrence, Jr. in Art Unit P/1776. Application was dated May 13, 2013.

j. 9,540,960 B2, "Low Energy Nuclear Thermoelectric System," January 10, 2017, by Nicolas Chauvin: discusses engineering design of a thermal generator to utilize transmutation reactions to produce heat for use in mobile applications; indicates use of heater and radio frequency energy to energize nickel powder in a reaction chamber and a shield to block any gamma rays emitted by the transmutations. Patent examiner: Jesse Bogue in Art Unit P/3748. Application was dated March 22, 2013.

k. 10,465,302 B2, "Modular Gaseous Electrolysis Apparatus with Actively-Cooled Header Module,

Co-Disposed Heat Exchanger Module, and Gas Manifold Modules Therefor," November 5, 2019, by Ernest Charles Alcaraz et al.: discusses design of a gas or gaseous type of cold fusion device with a cooled header for electric connections, a heated anode and co-disposed cathode within a reaction chamber, a heat exchanger to extract heat from the reaction chamber, gas source and collection manifolds, and an improved electronic control circuit. Patent examiner: Harry D. Wilkins, III in Art Unit 1794. Application was dated February 22, 2017.

l. 10,480,084 B1, "Modular Cooling Chamber for Manifold of Gaseous Electrolysis Apparatus with Helium Permeable Element Therefor," November 19, 2019, by Monte S. Chawla et al.: discusses design of an improved modular cooling chamber for reaction gas product collection manifolds of gas or gaseous cold fusion devices, with a helium permeable element to separate helium from hydrogen and/or deuterium reactants. The cooling chamber can provide a controllable thermal gradient across the helium permeable element to promote helium transport through the cooling chamber. Patent examiner: Jason M. Greene in Art Unit 2855. Application was dated March 3, 2017.

These examples of patents appear to fall into two categories: those more closely associated with fusion, deuterium, energy, and helium and those more closely associated with transmutation, hydrogen, energy, and transmutation products. All are related to cold fusion and LENR that was generally unheard of before Pons and Fleischmann in 1989. Patents (a–d), (f–i), and (k–l) are more closely associated with fusion, deuterium, energy, and helium, and patents (e) and (j) appear to be more closely related to transmutation, hydrogen, energy, and transmutation products.

Information to support an industrialization program can also be found in international patents and patent applications after they are published by the patent office. A long list of published applications is provided in patent 9,540,960 B2; at least five patents or patent applications related to cold fusion or LENR have been created by Nichenergy, a research company in Italy.[57] The Nichenergy patents appear to focus on transmutation, hydrogen, energy, and transmutation products.

6

ROBUST CATHODE DESIGN

THE LAST THIRTY YEARS OF COLD FUSION RESEARCH have demonstrated that cold fusion reactions are difficult to produce, at least consistently over a long period of time. When they occur, the conditions need to be replicated well for the reactions to continue. Earlier, some of the reasons were mentioned that liquid electrolysis experiments would be difficult to impossible to scale-up to greater power levels. Designs for gas or gaseous cold fusion generators also have their own set of issues.[58] For example, the anode and cathode must be sufficiently spaced apart to prevent high voltage breakdown. All possible means should be implemented to load the cathode, such as high gas pressure, thermal diffusion, and strong electric fields. The volume of cathode reaction material must be sized to contain enough reaction sites for the desired amount of power, taking into account many sites where reactions will not occur. In addition, the cathode must be able to be installed with ease and later replaced.

Technical Background

Consider deuterium and/or hydrogen gas between an anode and cathode. In a gas-based system, molecules of gas impacting the surface of the cathode will travel at high velocity related to temperature (thermal kinetic energy). Gas pressure on the surface of the cathode is due to the number of gas molecules and their kinetic energies. Average velocity of the molecules can be easily calculated if this were of interest. The average density of molecules at any instant can be calculated from the ideal gas law, $PV = n\,RT$, where n is the number of moles of gas (1 mole = 6.02×10^{23} molecules) and R is the universal gas constant (R = 0.082 liters-atmospheres/moles-°K). If the system were operated at 10 atmospheres of pressure and 456°K (183°C, the Debye temperature for nickel), then one liter of the gas would contain 1.6×10^{23} molecules. A volume of one cubic micron (10^{-15} liter) adjacent to the cathode's surface would contain 1.6×10^8 molecules. These will need to penetrate through the cathode's surface and flow into the reaction material's microscopic cracks, crevices, and defects to support the cold fusion reaction process. Higher gas pressure is expected to be required for operational devices.

Hydrogen/deuterium concentration and, consequently, reaction rate are not expected to be uniform across the volume of reaction material, even if steady state conditions were reached, but will be localized into a relatively narrow region of the reaction material. The localized concentration of hydrogen and deuterium is expected to move about in response to small changes from a steady state. This can be demonstrated in equations of hydrogen flux through the reaction material as the sum of a drift term due to electric field and a diffusive term caused by the temperature gradient.[59] Localization and movement of hydrogen and deuterium in the reaction material have important consequences. With regard to cathode design, a long cathode with a large surface area is much better than a thick one. Reaction material for the cold fusion generator discussed in the next chapter was fabricated by high-pressure consolidation of nickel metal particles. It is in the shape of an open, solid cylinder, 35 cm long, with an inner diameter of 7 cm, an outer diameter of 12 cm, and a volume of 2600

cm^3. The consolidated reaction material is shown to the lower left of the accompanying photograph, which also shows a reaction chamber and its top closure or header.

Due to the role of phonons, it is assumed that reaction material in the cathode must be heated above its Debye temperature (183°C) for consistent operation and that the cathode's operating point will need to be accurately controlled for the reactions to continue. The cathode can be heated above this temperature with a separate, built-in electric heater. The additional heat produced by cold fusion reactions will need to be removed in a regulated manner so that temperature internal to the cathode is not reduced below its operating point.

Deuterium-deuterium (d, d) cold fusion experiments have demonstrated that sufficient energy can be produced in the microscopic, local vicinity where reactions occur to melt the reaction material. This concern stems from the observation of "volcanoes" formed from melted metal on cathode (palladium) surfaces. The volcanoes have a diameter of a few tenths of a micron to tens of microns. Depths are about the same as the diameters. Temperature of the material would need to be raised by at least 1500 degrees for melting. Since the heat capacity of metals is 25 joules per mole per degree-Kelvin, about thirty reactions at 3–4 MeV each (a total of 100 MeV or 0.016 nanojoule) would provide enough energy to "melt" a million (10^6) atoms of the material.

Calculation of Energy Limitations

A cubic centimeter of nickel would contain 10^{22} atoms if the metal atoms were separated by approximately 5

Angstroms (5×10^{-8} cm). Instead of solid metal, consider a cubic centimeter of cathode reaction material produced, for example, by consolidating nickel metal particles. If a cubic centimeter contains a total of 2.7×10^{10} spaces between the metal particles (emulating cracks/crevices/defects), it would have about 3,000 spaces along each x, y, z dimension. As an estimate of energy that could be produced in each small space, assume that it could be loaded with 20,000 deuterium and hydrogen atoms, providing the possibility of 10,000 (p, d) cold fusion reactions, and assume that each of the reactions were able to produce 5 MeV of energy. Gamma radiation from the (p, d) cold fusion reactions would be adsorbed by the entire cold fusion generator mass, rather than within the local reaction sites. Since 1 MeV equals 1.6×10^{-13} joule, 10,000 reactions would produce 8×10^{-9} joule. This is only eight nanojoules—a very small amount of energy. A total of 2.7×10^{10} spaces, however, may be able to produce 216 joules/cm^3. If this energy were produced each second, then it would result in 216 watts of power for each cubic centimeter of reaction material, which is about the same power density as that produced by nuclear fission power plants.

Instead of thinking about the metal surface internal to the very small spaces, it is also possible to visualize the internal surface itself as connecting into many narrow, one-micron-long linear channels and defects branching off from the small spaces. Each of these more narrow, linear channels could contain several thousand deuterium and hydrogen atoms. Assume that each space branches off to ten linear channels or defects, and that each channel contains 2,000 deuterium atoms, providing the possibility of 1,000 cold fusion reactions. These reactions would produce 8×10^{-10} joule in each channel, 8×10^{-9} joule in 10 channels, or 216 joules in each cubic centimeter of the cathode.

Compare this to (d, d) reactions if only deuterium were used. Visualize the nickel surface internal to each of these very small spaces as having an internal circumference of approximately 3.9 microns (3.9×10^{-4} cm), an internal surface area of 4.8 square microns, a volume of one cubic micron, and a radius of 0.6 micron. The surface would contain about eight thousand metal atoms around a circumference and 2×10^{7} atoms around its internal surface. Also consider the number of metal atoms in an imaginary sphere centered on and surrounding the small space. If the sphere has a radius of 2 microns, for example, it would have a volume of 40 cubic microns. The volume of 40 cubic microns, less the volume of small space (1 cubic micron), would contain 3×10^{7} atoms. Since only thirty reactions at 3–4 MeV each (a total of 0.016 nanojoule) can provide enough energy to "melt" a million (10^{6}) atoms of the material, these atoms (3×10^{7} atoms) would be expected to melt if they absorbed energy from nine hundred cold fusion reactions.

The reader may be interested in reviewing the calculations to determine if smaller or larger metal particle sizes would improve the outcome. For extended operation, however, it seems that the amount of energy produced per reaction site must be controlled, especially when the focus is on (d, d) fusion. One way to do this, if only deuterium were used, would be to limit the quantity of deuterium gas provided to the sites, for example, down to about one hundred atoms. Power rating would be proportionally impacted. A larger cathode might be used (greater cost) to increase the number of reaction sites and maintain power rating. The best advantage, however, might be obtained by limiting the quantity of deuterium compared with hydrogen to enhance the relative number of (p, d) fusion reactions.

7

GENERATOR DESIGN

THE NORTHERN VIRGINIA TEAM USED INFORMATION discussed in earlier chapters, along with other technical sources, to design a cold fusion generator (known as the "Mk12.31") that would help to bridge the gaps between experimental laboratory systems and a commercially useful device. A power level of 200 kW was chosen to be of practical interest. Key components are: a reaction chamber (or reactor); a top closure or "header"; a heat exchanger/boiler; a gas handling system consisting of four gas manifolds; and an advanced electronic control subsystem.[60] The generator is designed to be small enough to fit into a vehicle's engine compartment. It is designed to use deuterium and hydrogen gas to produce deuteron-deuteron (d, d) and/or proton-deuteron (p, d) cold fusion reactions. The (p, d) reactions should occur more easily in a cold fusion environment than (d, d) reactions, but produce intense 5.5 MeV gamma radiation that needs to be absorbed by the generator's total mass.

The electrochemical process between the anode and cathode can be referred to as "gaseous electrolysis," as it entails pressurized deuterium and/or hydrogen gas, elevated temperatures, strong electric fields, and a mixture of molecules, ions, and electrons. It is concerned with mechanisms that form positive ions from deuterium and hydrogen gas molecules, their movement to the cathode, and cathodic interactions. Sufficiently low voltage and high gas pressure are used to prevent high-voltage breakdown or avalanche discharge through the gas. By comparison, liquid electrolysis in early cold fusion experiments involved a liquid electrolyte with D^+ and/or H^+ cations moving to and interacting with the cathode.

Supporting Missions to Outer Space

The heat exchanger/boiler can be replaced by rows of solid-state thermoelectric components surrounding the reactor along with external cooling fins/heat radiators to remove the heat. This configuration could support NASA missions to outer space. For example, electricity produced by the generator could be used to power ion thrusters and to separate deuterium gas from lunar water. The deuterium gas could then be used in generators for planetary exploration.

Reaction Chamber

The reaction chamber is designed to be large enough to contain an anode, an appropriately sized cathode surrounding the anode, a microwave loop antenna, a temperature sensor, and insulators and physical supports for these components. It can be fabricated from a large, type-316 stainless steel pipe or by modifying a commercially available reactor. Side ports connect to gas manifolds in the gas handing system. The microwave antenna is planned to be used during system start-up to jiggle electrons in the gas between the anode and cathode. An electric heater in the anode can be used to increase temperature of the inner cathode surface. The reaction chamber is designed to provide high pressure, temperature, and electric field conditions that facilitate hydrogen and deuterium absorption into the cathode's reaction material. Although pressurized hydrogen and deuterium are supplied to the reactor in the form of gas molecules, the region between the anode and cathode will contain elevated temperatures, a strong electric field, and a mixture of ions and electrons, as well as gas molecules. The reaction chamber should be able to operate with gas pressure as high as 1000 psi at 500°C, but early prototypes are expected to operate at 150 psi and 300°C. Pressure can be increased as needed to prevent high-voltage breakdown between the anode and cathode. The reaction chamber also must not leak with changes of pressure and temperature, such as during generator start-up and shutdown.

The reaction chamber's cylindrical configuration supports an even electric field impressed between the anode and cathode to coerce positive gas ions in the direction of reaction material in the cathode. Several ion-forming mechanisms are involved in the gas. The most important is due to collision of thermal electrons with gas molecules. The resulting mixture will contain different species of positive and negative ions and molecules interacting with various probabilities. Positive ions of hydrogen and deuterium can be repelled by the positive anode and accelerated toward the negative cathode at rates determined by their mass and electric charge. Further movement of deuterium and hydrogen though the reaction material after the ions enter its surface can be caused by the electric field and temperature gradient. Reaction chamber rough order of magnitude (ROM) cost: $75–100K.

The header is a separate but integral part of the reactor/ reaction chamber. It enables the generator to be opened and closed for service of internal components. The header is subjected to the same high temperatures and pressures as the reactor and must not leak during long periods of operation. It may be fabricated from type-316 stainless steel. Design of the header includes rugged attachments to the anode, microwave loop antenna, temperature sensor, and to their insulators and physical supports. Due to its size and weight (approximately sixty pounds), the cathode is a separate component that is not attached to the header. It is electrically grounded to the reaction chamber. The header is also designed to cool high-pressure feedthroughs that connect electrical components in the reaction chamber with the electronic control subsystem. A cooling manifold is used to cool a metal plate through which the feedthroughs can be mounted. The design also includes a gas safety cover surrounding the cooling manifold to detect any possible leak. Header ROM cost: $60–80K.

Heat Exchanger/Boiler

The heat exchanger is designed to rapidly remove heat, but on a controlled basis, from the outer surface of the reaction chamber so as to enhance thermal diffusion through the cathode and reaction chamber wall. It is a low-volume, quick-acting, flash boiler that is mounted to and surrounds the reaction chamber, and should also be fabricated from type-316 stainless steel. A mist of water or other coolant is provided through high-pressure spray nozzles mounted on its outer wall and directed toward the surface of the reaction chamber (see drawing). The heat exchanger is also designed to supply steam in a closed-loop configuration to downstream electric turbines or generators.

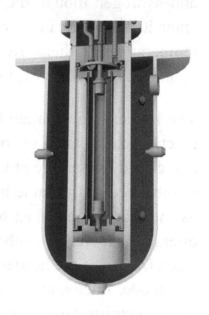

The reaction chamber and heat exchanger are designed to be sufficiently thick to absorb practically all 5.5 MeV gamma radiation from the (p, d) reactions, converting the radiation into heat. The XCOM photon-cross-section database available from the National Institutes of Standards and Technology (NIST) can be referenced to determine amounts of gamma ray energy absorbed in various materials, and attenuation calculations can be performed on the internet with one of several x-ray/gamma radiation calculators. The example in the below chart shows that, while only 6 percent of the energy from 5.5 MeV (p, d) reactions would be stopped by 2.5 mm, about 71 percent will be attenuated in 5 cm. This is about the distance through the cathode, a steel sleeve around the cathode, the reaction chamber wall, and the heat exchanger/boiler wall. Gamma radiation will be absorbed through a combination of the photoelectric effect, Compton scattering and pair production. Additional radiation could be absorbed if the steel sleeve around the cathode were replaced by a tungsten sleeve. Any remaining radiation could be prevented from being hazardous to operators by housing the generator in a room that provides a safe keep-out distance.

Gamma Radiation Attenuation through Reaction Chamber and Heat Exchanger

Effect	Attenuation (%)								
	Reaction Material			Sleeve	Reaction Chamber Wall		Boiler Wall		
Depth (cm)	0.25	0.5	1.0	1.5	2.0	3.0	4.0	5.0	10.0
Photoelectric	0.005	0.01	0.02	0.03	0.03	0.05	0.05	0.6	0.8
Compton	4.44	8.6	16.2	23.0	28.9	38.8	46.5	52.3	67.8
Pair Production	1.5	3.0	5.6	8.0	10.0	13.5	16.1	18.1	23.4
TOTAL	6.0	11.6	22.0	31.0	39.0	52.4	63.0	71.0	91.6

The amount of radiation leakage during operation will need to be monitored with a high-energy gamma radiation detector or spectrometer located outside of the header or heat exchanger. The spectrometer should be able to provide a history of radiation levels during system operation, showing the radiation's full-energy peak at 5.5 MeV, with related lower energies. The appearance of the spectrum can be estimated with the Gamma Spectrum Generator (GSG) provided by the Joint Research Centre Institute for Transuranium Elements.[61] Heat exchanger ROM cost: $30–40K.

Gas Handling System

The gas handling system includes four gas manifolds connected to the reaction chamber. The manifolds are constructed with high pressure-rated, stainless steel tubing and fittings. The design minimizes gas volumes external to the reaction chamber and enables small quantities of supply gas to be determined through pressure, temperature, and volume calculations. Each manifold includes gas measurement chambers, temperature and pressure sensors, and mechanical and electric valves to control gas flow. To limit the amount of heat from the reaction chamber and protect the sensors and valves from high temperature, each manifold contains its own cooling chamber. The electrical components are powered and regulated by the electronic control subsystem. One of the gas manifolds provides controlled amounts of high-pressure hydrogen and/or deuterium gas to the reaction chamber. Only about 2×10^{17} molecules of gas per second are needed for 200 kW of power. Delivery of this small amount of gas, therefore, must be provided by small, high-pressure gas puffs. This is unusual in that it is only a millionth of the quantity of gas ordinarily dealt with in conventional power and heating systems. Another gas manifold provides argon carrier gas to the reaction chamber for start-up and maintenance. A separate safety tank is used to capture hot, high-pressure gas from any hazardous overpressure condition.

Two gas manifolds are designed to remove gas from the system. One of the two manifolds is designed to remove gas from the entire system during maintenance. The other manifold is designed to separate the helium gas by-product, to quantify the amount of helium separated, and to store it for later extraction and analysis. The design employs diffusion through a thin membrane to separate helium from deuterium and hydrogen.[62] Materials considered for the helium permeable element, such as zirconia, fused silica, and silica glass, are able to withstand high temperatures. Sufficiently high helium diffusion rates are possible due to helium's small, monoatomic molecule diameter compared with hydrogen's larger diatomic molecule.

About 2×10^{17} helium atoms can be anticipated to be produced per second from 200 kW. These will need to be removed periodically so that additional hydrogen and/or deuterium gas can be added, enabling the generator to operate continually for long periods. A year of continuous operation at 2×10^{17} nuclear reactions per second can produce 6×10^{24} helium atoms (10 moles) that occupy about 200 liters at standard temperature and pressure. A capability to remove incremental and predetermined quantities of helium can also help balance pressure-related, variable operating conditions within the reactor. Additionally, helium is an irreplaceable natural resource of limited extent. Collection and storage of helium can result in a profitable resource due to its commercial uses.

The design of this gas manifold also includes a measurement chamber with a special type of electronic interface for matter output (EIMO) used to determine the amounts of separated helium by-product that are temporarily stored. In addition to very sensitive pressure and temperature sensors, this component includes an acoustic sensor to measure sound velocity through the collected gas. The combination of sound velocity, temperature, and pressure can then be used to estimate helium quantities. Temporarily stored gas can subsequently be extracted from the manifold and evaluated off-line with commercial binary gas analyzers. Gas handling system ROM cost: $80–100K.

Electronic Control Subsystem

The electronic control subsystem is designed to provide electric power to electronic components in the gas handling system and reaction chamber and to regulate, record, analyze, and control generator operations. The design includes a robust uninterruptable power supply (UPS), separate power supplies for the electronic components, and computers/controllers that use data from these components to determine steps required for generator operation. Combined voltage/current, variable power supplies are provided for the anode, anode heater, and microwave initiator. The electronic control subsystem is also designed to communicate digitally with a broad number and type of electronic components, to include the electronically controllable gas valves and sensors that provide temperature, pressure, acoustic, electric current, and nuclear radiation data.

Examples of electronic control subsystem functions include real-time, remote setup and control of generator operations through video screens; automated start-up and fail-safe shutdown; continuous monitoring of pressures and temperatures to maintain safe operating conditions; detection and alerting of significant events; controlling the small increments of high-pressure hydrogen and deuterium gas into the reaction chamber; facilitating diffusion of hydrogen and/or deuterium through the cathode; automatically adjusting system operational parameters; and determining when maintenance is required. Electronic control subsystem ROM cost: $60–80K.

The ROM cost for miscellaneous parts, components and supplies: $175–250K.

The generator will have near-term practical applications in laboratory cold fusion testing activities, such as the study of gas electrolysis; cooling and heat transfer with the heat exchanger; heat extraction to provide useful output; and feedthrough cooling to enable longer periods of operation between maintenance. The system is designed to be used eventually as the prime power source in community-based power plants. All components and parts can be manufactured using standard production methods. Additional technical information is provided in the appendix.

8

STEPS FOR ADVANCED DEVELOPMENT AND DEMONSTRATION

O NE OF THE FIRST STEPS IN AN ADVANCED DEVELOP-ment and demonstration program will be to compare designs being pursued by others in the cold fusion community (and industry, if possible) with those of the Mk12.31. Due to urgency of the present climate crisis, it is important to consolidate physics and chemistry under-standings from the various designs into the program's most promising design.

For comparison, the key attributes for the Mk12.31 design discussed in this anthology are as follows:

- The generator uses deuterium and hydrogen gas to produce deuteron-deuteron (d, d) and proton-deuteron (p, d) cold fusion reactions. Proton-deuteron (p, d) fusion should occur more easily in a cold fusion environment than (d, d) fusion.

- Due to the difficulty in loading gas into cathode reaction material, loading involves a combination of high gas pressure, thermal gradient, and high-voltage electric field. A heat exchanger/boiler facilitates the thermal gradient and contains spray nozzles for quick response to changes in reaction chamber temperature and to remove heat from the system.

- Designed to produce 200 kilowatts (kW) of heat, which can be converted into mechanical horsepower for community-based power plants. Each (p, d) fusion will produce an atom of helium-3 and 5.5 MeV of energy from gamma radiation. About 2×10^{17} reactions per second will produce 200 kW.

- The design is supported by previous research showing that (p, d) fusion produces helium-3; that (p, d) fusion in a cold fusion environment should be easier than (d, d) fusion; and that deuterium ice experiments with protons produced helium-3.

- Expected to operate continually for longer periods than systems producing energy by transmutation or only (d, d) fusion where their cathodes would be degraded more rapidly.

- Gamma radiation energy is to be absorbed and contained by mass of the system's physical components.

- Energy produced by each (p, d) fusion is about the same as energy that could be produced by averaging the different types of (d, d) fusion. Less deuterium will be needed.

- The design includes four gas handling manifolds and a sophisticated electronic subsystem to monitor and control system operation.
- Deuterium and hydrogen are provided in small, high-pressure gas puffs that contain only a millionth of the quantity of gas ordinarily dealt with in conventional power systems.
- The design includes a method to extract helium by-product gas, enabling addition of deuterium/hydrogen for long-period operation.

The next steps will need to develop an even more detailed understanding, requiring scientists and engineers with advanced knowledge of physics and engineering and who are highly experienced in working in R&D laboratories, team oriented, and committed to further innovation. Industrial partners will need to be technically advanced R&D companies.

Estimated Program Cost

At least five years of work and $25 million will be needed to obtain an industrial prototype. Due to its technical complexity, however, the advanced development and demonstration program could realistically cost five to ten times this amount.

Program Activities

Steps for the program may be divided into phases, such as: test planning and preparation; laboratory testing in a relevant environment; prototype demonstration; validation and verification (V&V); and preparation for production. The first three years of the program should emphasize laboratory demonstration and testing, while the fourth and fifth years should focus upon V&V and manufacturing.

Test Planning and Preparation

Year 1 (e.g., 2021) will need to include technical, administrative, human resource, and financial tasks to prepare for later steps in the program. Examples of tasks are as follows:

a. Confirm financial support and allocate funding to work breakdown schedule (WBS) tasks.

b. Obtain commitment from R&D companies comprised of senior scientists and system engineers who are proficient in understanding the physics and engineering involved, building test equipment, and performing experimental testing. Dedicated laboratory space and at least ten full-time scientists/engineers will be required.

c. Detailed team review and discussion of the knowledge base, physics, and engineering involved in the testing activities and the technical approach for the five-year program. Includes all relevant topics, such as hydrogen thyratron tube design, gas/gaseous electrolysis, cathode operating parameters, and regulation of remote electronic components.

d. Physical layout developed for generator and laboratory analytical equipment.

e. Equipment procurement.

f. Develop design for physical safety enclosure to protect workers, taking radiation containment into account.

g. Evaluate permeability and absorption of deuterium into consolidated cathode material.

h. Team discussion of possibilities for spin-off/novel intellectual property.

i. Safety review and procedures documented to prevent physical, high voltage and other hazards.

j. Senior members of scientific community (i.e., three to four) selected for annual peer review.

k. Key physical processes modeled, for example, using COMSOL Multi-physics simulation series of computer codes.

l. Initiate search for public or public-private financial support needed in final two years.

m. Installation of generator subsystems/components.

n. Operational steps in electronic control system software reviewed and modified as necessary.

o. Laboratory analytical equipment installed, such as calorimeters, helium mass spectrometer, high-energy gamma ray spectrometer, midenergy gamma ray spectrometer, high-energy Geiger-Muller detector, radiographic detectors (CR-39/Lexan and film), ion microprobe, and data logging system.

p. Team discussion of applicable rapid prototyping principles for equipment to be developed.

q. Preliminary operational testing of assembled subsystems (without hydrogen or deuterium). Documentation of test configuration and results.

r. Checkout and baseline calibrations performed on analytical equipment.

s. Peer review committee documentation of progress in first year, including comparisons to progress by others reported in the literature and in patents.

t. Discussion of strengths and deficiencies in test design and procedures related to a specific energy application.

u. Program status briefed/presented to supporting philanthropists, agencies, and organizations.

Laboratory Testing in a Relevant Environment

Year 2 (2022) should focus upon experimental testing in a specified, realistic environment. This will include tasks to reach Technology Readiness Level-5 (TRL-5) and to prepare for prototype hardware demonstration in the third year of the program. Examples of tasks are as follows:

a. Safety enclosure constructed and installed.

b. Team discussion of laboratory safety procedures.

c. Team completes checkout of test equipment and begins active testing.

d. Team determines the amount of chemical energy produced versus energy from nuclear reactions.

e. Amounts of by-product gas (helium-3 and helium-4) determined and correlated with heat produced.

f. Study of related research worldwide continued with incorporation of applicable new information into the testing program.

g. Discussion of novel technical information derived from testing program and possible spin-off inventions and patents.

h. Modification of test equipment as needed to resolve deficiencies determined in test period.

i. Steps in electronic control system software checked to ensure proper operation.

j. Procurement and installation of additional laboratory analytical equipment as needed.

k. Calibration checks performed on analytical equipment.

l. Continuation of effort to establish public-private sources of support.

m. Peer review committee assessment.

n. Program status briefing for supporting philanthropists, agencies, and organizations.

Prototype Demonstration

Year 3 (2023) will need to focus on prototype hardware demonstration. This will include additional tasks to reach TRL-6 and to prepare for prototype validation and verification in the fourth and fifth years of the program. Examples of tasks include:

a. High-purity gases (hydrogen, deuterium, argon) obtained from suppliers.

b. Team review of laboratory safety procedures.

c. Team testing of demonstration system and documentation of test configuration and results.

d. Samples of collected helium gas correlated with heat produced.

e. Team discussion of energy production at levels of practical interest.

f. System configuration evaluated for planned power application.

g. Acquisition of improved cathodes.

h. Team efforts to ensure that supporting elements of the system are realistically demonstrated.

i. Draft radiation standard developed and discussed with DOE and the NRC.

j. Team determination that performance objectives in an operational environment can be met.

k. Team ensures that information presented to the scientific community is accurate.

l. Component-level specifications and drawings developed.

m. Manufacturing company(ies) selected to observe validation and verification.

n. Peer review committee assessment.

o. Sources of financial support identified for fourth and fifth years. Funding allocated to WBS tasks.

p. Plans developed to patent new intellectual property.

q. Progress briefed to supporting philanthropists, agencies, and organizations.

Validation and Verification

Year 4 (2024) should focus upon V&V of technical data. Elements of technical information needed to support the transition between laboratory cold fusion experiments and manufacturing a commercially useful device will be developed. Activities will include discussions with engineers from interested manufacturing company(ies) who will witness testing and determine how well the generator's design and specifications address energy requirements. Year 4 will also be used to plan production activities. Examples of tasks are as follows:

a. Procurement and installation of new or redesigned equipment and software for use in V&V testing.

b. High-purity gases (hydrogen, deuterium, argon) obtained from suppliers.

c. Laboratory safety procedures reviewed by all personnel.

d. Team testing of new/redesigned system with active gases and documentation of test configuration and results.

e. Samples of collected by-product gas correlated with heat produced.

f. Team comparison of data from validation testing with data obtained in previous experimental tests.

g. Team reviews steps in the operational software.

h. Patent applications developed and submitted to USPTO.

i. Licenses issued to manufacturing company(ies).

j. Draft developed for a maintenance and supply plan.

k. Draft of radiation standard finalized and submitted to DOE and the NRC.

l. Documentation of applicable rapid prototyping methods.

m. Specifications and drawings reviewed with manufacturing companies. Required modifications determined, to ensure device can be produced by industry. Specifications and drawings finalized.

n. Discussions with utility companies on test results and applications in community-based power plants.

o. Peer review committee (final) assessment.

p. Progress briefed to supporting philanthropists, agencies, and organizations.

Preparation for Production

Year 5 (2025) should continue to focus on data V&V and preparation for manufacturing/production. It will ensure that the specifications, drawings, and other documentation adequately describe the system. Examples of tasks are as follows:

a. Laboratory safety procedures reviewed by all personnel. Safety documentation provided to manufacturing companies.

b. Cathodes purchased (as needed) to develop additional verification data.

c. System testing (as needed) to develop additional verification data.

d. Samples of by-product gas collected and analyzed as needed.

e. Software specifications and coding logic are compared against steps required for accurate, remote operation.

f. Patent applications revised and resubmitted to USPTO.

g. Lists of applicable manufacturing and operating standards developed by manufacturing companies and discussed with team.

h. System design data transitioned to production facilities.

i. Team publishes technical reports on physics and engineering accomplishments.

j. Progress briefed to supporting philanthropists, agencies, and organizations.

9

DEMONSTRATION EXPERIMENT

ENGINEERS ASSIGNED THE TASK OF DEVELOPING A small cold fusion generator demonstration experiment will face complex and possibly unsolvable technical issues related to size restrictions. A useful operational unit providing less than 200 kW (heat) is not expected to be able to be built and demonstrated. But if successful, a goal for such demonstration experiment should be to produce energy for some useful purpose, such as a remote, unattended power source or a device to complement spaceborne radioisotope thermoelectric generators (RTGs).

Design of a 10 kW Demonstration

Since the efficiency for converting heat to electricity is less than 50 percent, about 10 kW of heat will be needed for 5 kW of electricity. The volume and dimensions for the cathode could be estimated from this, along with the required rate and quantity of deuterium. As few as 10^{16}

atoms reacting per second should produce about 10 kW. The number of extremely small, linear defects, cracks, and crevices required in the cathode's reaction material can be determined, along with the amount of heat required to raise the cathode to its Debye temperature. From these data, it would be possible to determine the electric field strength, required separation between the anode and cathode, and gas pressure to prevent high-voltage breakdown. The cathode, anode, and heater could then be designed.

An amalgamation of these ideas was used for the example shown in the accompanying drawing. The cathode is to be made of consolidated nickel powder containing a great many

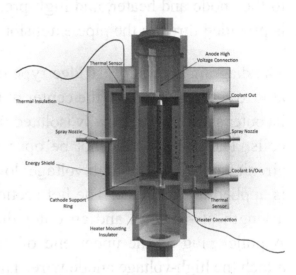

microscopic-sized cracks and crevices. The heater also serves as the anode. Very high gas pressure is used to prevent high-voltage breakdown. The reader may be interested in comparing these features with other designs on the internet by S. Focardi et al., Defkalion Green Technologies, A. Rossi, and Gyorgy Egely, for example.[63, 64, 65, 66]

Construction

The cathode where the reactions are to occur has an approximate volume of 200 cm^3, is 4 inches long, has an inner diameter of 1.5 inches and an outer diameter of 2.5 inches. It is inside a stainless steel pipe that is 8 inches long, has an inner diameter of 2.5 inches and an outer diameter of 2.906 inches. Each end of the pipe is threaded so as to be joined to pipe extensions (not shown) used for gas and electrical inputs. Electrical connections are made to the anode and heater, and high-pressure deuterium gas is provided through the pipe extensions.

The diagram shows a cartridge type of heater that is also able to serve as anode in the center of the cathode because its outer sheath is electrically isolated from internal heating coils. The heater would not be operated, and its electric circuit isolated, while high voltage, low current electricity is applied to the anode. The hot section of the sheath has a length of 4 inches and an outer diameter of 0.5 inch. A puller plug on the upper end of the sheath is used to attach the high-voltage anode wire. The anode wire passes through a ring insulator above the heater. On the other end is an inch-long, unheated section and flat flange for mounting the heater. Power for the heater is provided by two wires from this end. The flange is mounted to another ring insulator that contains holes so that gas can flow into the space between the cathode and anode/heater. The manner in which these insulators are secured to the surrounding pipe is not shown. Additional ring insulators are used to support the cathode.

The outer part of the device consists of a flow calorimeter with an internal fluid volume of approximately 2 liters. It is made from a 6-inch-long stainless steel pipe with an inner diameter of 6 inches and outer diameter of 6.56 inches. The top and bottom of the calorimeter are made of stainless steel disks welded to the ends of the pipe and to the inner pipe, while still exposing threads to mate with the pipe extensions. The calorimeter contains spray nozzles that are exercised during start-up and later as needed for cooling. It contains thermal sensors and inlet and outlet pipes for adding and removing water.

A tight fitting, high-Z tungsten metal energy shield and thermal insulation fabric are added around the calorimeter. The energy shield converts some of the gamma radiation produced in the cathode into heat adsorbed in the calorimeter. The thermal insulation helps to retain heat and increase accuracy of thermal measurements.

Assembly and Operation

Before installation of the cathode, heater, and insulators, great care is required to cleanse the internal metal body of the device and pipe extensions of residue and organics used in manufacturing and assembly of the parts. Steps of inspection also ensure that the cathode, heater, and insulators are not contaminated. Parts must subsequently be handled with clean gloves. After inspection, the gas supplies and vacuum system are connected to ports in the pipe extensions. Thermal and pressure sensors and power for the electric heater and high-voltage anode source are connected. Control and measurement software would be exercised to demonstrate that operations can be performed remotely. The computer/data acquisition system should be set to record data continuously (e.g., each second) from current and voltage sources and temperature and pressure sensors.

Subsequent steps should be at a safe distance and behind a suitable safety shield. A vacuum is pulled on the system, and power for the heater is turned on in an attempt to remove oxygen from system components. The device should be subjected to a sequence of vacuum and high-temperature cycles, and allowed to bake out until no further pressure changes occur. Care is needed to limit heater power to less than its maximum operating power rating. The sequence of steps from this point varies according to experimental objectives. After power is removed from the heater and the device is allowed to cool, pressurized deuterium gas is added and pressure readings are recorded over time to check for leaks. If no leaks are found, additional deuterium can be added to reach approximately half of its operating pressure. Pressure readings are again recorded to check for leaks. Readings are also monitored to determine the amount of cathode loading with deuterium gas, and loading steps are repeated as needed. After sufficient loading, the cathode would again be subjected to high temperature from the heater. The spray nozzles would be cycled briefly to cool the calorimeter's inner surface and hence the outer part of the adjacent cathode. These loading steps are repeated as additional deuterium is loaded into the cathode. Water flow through the calorimeter is established, and the rate is adjusted to ensure that the adjacent cathode is not cooled below its Debye temperature. At this point, the amount of heat generated in the cathode can be determined with the calorimeter. Small samples of reaction gas can be extracted through a port in the pipe extensions and subjected to mass spectrograph analysis.

10

BENEFITS OF COLD FUSION TECHNOLOGY

OLD FUSION SHOULD BE IMPLEMENTED AS THE NEW, key source of power as the world moves to sustainable, carbon-free energy, although this will be difficult to achieve, requiring industry and government leadership. Deuterium required can be extracted from water. As little as 0.0001 percent of the deuterium in global water can provide 10^{25} joules of energy.

Global Water: 1.3×10^{18} m^3
[H/D=O] Water: 4.1×10^{14} m^3
Energy from 0.0001%: 10^{25} joules

This amount of deuterium is sufficient to power the 800 million households on earth and meet the needs of industry for hundreds of years. The large amount of deuterium remaining could supply humanity's energy needs for thousands of additional years.

Deuterium Cost

Deuterium has many applications due to its physical and nuclear properties. One of the largest uses is in the form of heavy water as a neutron moderator for some types of nuclear fission reactors. It is also used in manufacturing semiconductors and optical fibers. Cold, liquid deuterium (D_2) is used as a moderator in neutron scattering experiments. Since deuterium behaves like hydrogen (protium) chemically, it is used in biochemistry as a nonradioactive tracer. Deuterium along with tritium is also planned to be used as a fuel in hot fusion reactors.

Production of heavy water and deuterium has been an important industrial process since the early 1940s. This involves extraction of heavy water (D_2O) from water by a chemical exchange process or by fractional distillation. Deuterium gas is separated from the heavy water by electrolysis. Heavy water is produced in several countries, mainly for use in nuclear reactors. It is also widely exported. The World Integrated Trade Solution database, supported by the World Bank in collaboration with the United Nations and World Trade Organization, indicates that 106,285 kilograms of heavy water with a value of $28 million were imported into the US in 2018.[67] Heavy water's cost can be estimated from this to be about $300 per kilogram. Heavy water has a mass of 20.03 grams per mole, and a kilogram contains 50 moles (1 mole contains 6.02×10^{23} molecules). The estimated cost per mole is about six dollars.

The cost of deuterium gas varies widely, depending upon suppliers, purity, and volume. Three nine's purity (99.9 percent) is less expensive than gas with greater purity; high-volume purchases are less expensive than small volumes. The density of deuterium gas is 4.028 grams per mole, or 0.179 grams/liter (4.028 grams per 22.4 liters) at 25°C and standard room temperature. Recent research has shown that zeolite or nanoporous sieves may be used to separate deuterium from a mixture of hydrogen and deuterium, possibly further reducing its cost.[68, 69] The use analysis below assumes, however, that the future cost of deuterium gas will be about ten dollars per mole.

Community-Based Power Plants

Initial implementation of cold fusion technology is envisioned to be for community-based power plants or "microgrids" that supply electricity to local buildings and small communities. Individuals and communities worldwide are interested in local power that is independent of long-distance utility power grids, that reduces their power cost, or that remains resilient in the face of natural disaster.

Gas and fuel oil–powered generators have previously been used to supplement grid-based power. Homeowners and communities are now transitioning to wind and solar power due to the decreasing cost of solar cells. Battery backup can be used if a steady stream of energy is needed over a short period. These local power systems can also be connected to the main power grid so that they return electricity to the grid for use by others. Supporting technical information and training courses for these systems are on the internet.[70]

Local power systems are also in the planning stage by many communities across the globe. Fire departments in California have begun installing their own microgrids to reduce the effects of damage to utility lines from forest fires that make communications difficult in areas needing support. Boston has issued a plea for neighborhood microgrid networks as a standard part of its climate action plan. Other cities are following suit in an effort to diminish climate change and environmental pollution. Community-based power and microgrids are useful in areas where long-distant utility power is not readily available and/or reliable, such as the following:

- undeveloped and developing countries
- new "clean" communities developed in previously outlying areas
- expansive areas where relatively small and dispersed populations live far from built-up areas of existing power grids
- large farms or ranches with limited or no access to other forms of power
- highly productive and less costly vertically oriented farming

Local power systems are also important when lower cost and critical backup power are required in large hospital systems, large school systems, industrial plants, and critical government facilities. The US Department of Defense uses local power systems for strategic facility support on bases throughout the country and has dedicated additional resources for renewable systems to reduce the effects of climate change. Energy from cold fusion generators can be essential in these types of electric power applications.

Future Use Cost Analysis

Consider a 200 kW generator that converts its heat to electricity with 50 percent efficiency and produces 100 kW of electric power. A typical residence requires about 4–5 kW. If a 200 kW unit were used to provide 100 kW of electricity, it could service twenty homes in a local community.

Assume that a small, one-story building in the local community housing the generator plus cabling to individual residences will cost $500K. The cost for prototype demonstration hardware was estimated earlier to be between $480K and $650K. First production unit (FPU) hardware cost could be half of $650K for industrial consumers or $325K. Cost for later production units might be further reduced through technical innovation. Each of the twenty homes in the community would be responsible for $41K of the $825K building and hardware cost. If the system has a thirty-year operational lifetime, annual building and hardware cost would be about $1,367 per residential unit.

The system is sufficiently complex to require remote operation and periodic maintenance by a local utility company. The cost for gases (deuterium, hydrogen, argon) is estimated to be $200 per year. Only 10 moles of deuterium ($100) would be required annually. Cathode replacement is expected to cost $25K but prorated as $5K per year. Utility company cost for operation and maintenance, including these expenses, is anticipated to be about $25K per year. Each residence would be responsible for $1,250 annually.

Annual cost of electricity for each residential unit can be estimated to be about $2,617 ($1367 + $1250). With 8,760 hours in a year, the cost of electricity would be thirty cents per hour for 5 kW, or about six cents per kW-hour. This is much lower than today's cost of electricity from the power grid (e.g., ten to twelve cents in Virginia; twenty-two cents in Alaska; and thirty-two cents in Hawaii). Waste heat from the system could also be made available at no or low cost for other heating applications. Exemption from future global pollution taxes would be expected, since the cold fusion system should significantly contribute to conserving natural resources.

Conclusions

Development of cold fusion technology will result in important benefits for humanity. Examples of specific benefits include the following:

- Potential for thousands of times more energy than contained in the same volume of gasoline.
- Clean and abundant fuel.

- Radiation produced during operation can be prevented from being hazardous by easily-provided container and safe keep-out distance.
- Radiation not emitted when generator is not operating.
- Possible power source for electric vehicle refueling stations.
- No emission of carbon dioxide (CO_2) into the atmosphere.
- Energy sufficient to power small communities.
- Local power generation facilities will reduce reliance on conventional power distribution systems.
- Maintenance costs and physical vulnerability of electric power distribution grid significantly reduced.
- Thermal energy returns 50–100X input energy.
- Systems are relatively small/compact.
- Little risk to operators provided reasonable safeguards are in place.
- Does not use radioactive tritium.
- Strong commitment and rapid development will ease concerns about global warming and climate change.

Cold fusion energy is produced from hydrogen and deuterium, which are abundant, low-cost resources. It is a clean form of energy. No radioactive materials are used. The process does not emit carbon dioxide or other harmful gases. Materials used in replaceable cathodes are abundant natural resources. Spent cathodes can be recycled when they are no longer usable, and their materials reprocessed into new cathodes. Reprocessing will produce little to no hazardous materials. Also, cold fusion generator development will support new industries that benefit the economy. Manufacturing, installation, and maintenance will result in new jobs and sources of revenue. Cold fusion can also reduce energy costs for homes and businesses.

Contributions from local engineering companies, program advisors, and Northern Virginia technical team members on the Mk12.31 are greatly appreciated. Team members included Ernie Alcaraz, Monte Chawla, Roy Collette, Randy Davis, Dorin Jannotta, Austin Lowrey, Tom McGraw, Fred Sandel, Don Waltman, and Rick Woll.

REFERENCES

Note. The indicated web addresses were accessed on October 12, 2020. Due to the dynamic nature of the internet, however, the addresses or links may change subsequent to publication.

1   Martin Fleischmann and Stanley Pons, "Electrochemically Induced Nuclear Fusion of Deuterium," submitted to the *Journal of Electroanalytical Chemistry* (March 11, 1989).

2   Pamela A. Mosier-Boss et al., "Investigation of Nano-Nuclear Reactions in Condensed Matter," report from the Defense Threat Reduction Agency (2016).

3   Website at https://iscmns.org/.

4   Website at https://lenr-canr.org.

5   See "DOE Review," https://lenr-canr.org/wordpress/?page_id=455.

6   David J. Nagel, "Evidence of Operability and Utility from Low Energy Nuclear Reaction Experiments," NUCAT Energy LLC, Report 2017-01, August 1, 2017, https://www.lenr-canr.org/acrobat/NagelDJevidenceof.pdf.

7   Informational video "Chemistry 10.4 Energy and Chemical Bonds (Entahlpy)," Issacs TEACH, March 13, 2012, https://www.youtube.com/watch?v=DUtUNm5uSH4.

8   Informational video, "Bond Length and Bond Energy,", Chemical Essentials Video 52, Bozeman Science, December 17, 2013, https://www.youtube.com/watch?v=I9jd1Ew_YGU .

9   Informational video, "Nuclear Physics: Crash Course Physics #45," PBS Digital Studios, March 20, 2017, https://www.youtube.com/watch?v=lUhJL7o6_cA.

10 Informational video,"Nuclear Fission and Nuclear Fusion – What Exactly Happens in These Processes?", GRS Deutschland, December 10, 2015, https://www.youtube.com/watch?v=xrk7Mt2fx6Y.

11 Bruce M. Steinetz et al., "Novel Nuclear Reactions Observed in Bremsstrahlung-Irradiated Deuterated Metals," NASA/TP-20205001616 (June 2020).

12 W. A. Fowler et al., "Investigations of the Capture of Protons and Deuterons by Deuterons," *Physical Review* 76, no. 12 (December 15, 1949): 1767–68 (see second paragraph on page 1767).

13 W. F. Hornyak et al., "Energy Levels of Light Nuclei. III," *Reviews of Modern Physics* 22, no. 4 (October 1950): 291–372 (see section L on page 299).

14 L. A. Radicati, "The Influence of Charge Independence of Nuclear Forces on Electromagnetic Transitions," *Physical Review* 87 (1952): 521.

15 Murry Gell-Mann and Valentine L. Telegdi, "Consequences of Charge Independence for Nuclear Reaction Involving Photons," *Physical Review* 91, no. 1 (July 1, 1953): 169–74.

16 M. H. Miles et al., "Correlation of Excess Power and Helium Production during D_2O and H_2O Electrolysis using Palladium Cathodes," *Journal of Electroanalytical Chemistry* 346 (1993): 99.

17 M. H. Miles, "Correlation of Excess Enthalpy and Helium-4 Production: A Review" (10th International Conference on Cold Fusion 2003).

18 Dimiter Alexandrov, "Cold Fusion Synthesis of Helium Isotopes in Interaction of Deuterium and of Hydrogen Nuclei with Metals" (2019 Cold Fusion Colloquium held at the Massachusetts Institute of Technology, March 23–24, 2019).

19 D. H. Wilkinson, "A Source of Plane-polarized Gamma-rays of Variable Energy above 5.5 MeV," *Philosophical Magazine* 43 (June 1952): 659.

20 J. Chadwick and M. Goldhaber, "A Nuclear Photo-Effect: Disintegration of the Diplon by Gamma-Rays," *Nature* 134 (1934): 237–38.

21 J. Chadwick and M. Goldhaber, "The Nuclear Photoelectric Effect," *Proceedings of the Royal Society, A* 151 (1935): 479–93.

22 H. Bethe and R. Peierls, "Quantum Theory of the Diplon," *Proceedings of the Royal Society A* 148, no. 863 (1935): 146–56.

23 "Handbook on Photonuclear Data for Application: Cross-Sections and Spectra," IAEA-TECDOC-1178, International Atomic Energy Agency (2000).

24 Frank Close, "Too Hot to Handle: The Race for Cold Fusion," Princeton Legacy Library (1991) (see page 312).

25 H. M. Taylor and N. F. Mott, "A Theory of the Internal Conversion of Gamma Rays," *Royal Society* 138, no. 836 (1932): 665–95.

26 Walter E. Meyerhof, "Elements of Nuclear Physics," McGraw-Hill (1967) (see pages 122–35).

27 M. E. Rose, "Internal Pair Formation," *Physical Review*, 76, (1949): 678.

28 S. Devons et al., "Life-time for Pair Emission by Spherically Symmetrical Excited State of the O16 Nucleus," *Nature*, 164 (1949): 586–87.

29 S. Devon and G. R. Lindsey, "Electron Pair Creation by a Spherically Symmetrical Field," *Nature*, 164 (1949): 539–40.

30 T. Mizuno et al., "Anomalous Isotopic Distribution in Palladium Cathode after Electrolysis," *Journal of New Energy*, 1(2) (1996): 37–44.

31 G. Miley and J. Patterson, "Nuclear Transmutations in Thin-Film Nickel Coatings Undergoing Electrolysis," *Infinite Energy*, 2(9) (1996): 19–32.

32 T. Mizuno, "Nuclear Transmutation: the Reality of Cold Fusion," Infinite Energy Press (1996).

33 David L Chandler, "Explained: Phonons," MIT News, Massachusetts Institute of Technology, July 8, 2010, http://news.mit.edu/2010/explained-phonons-0706.

34 Informational video, G. Rangarajan, "Mod-01 Lec-30 Cooper Pairs," Indian Institute of Technology Madras, July 24, 2013, https://www.youtube.com/watch?v=qTzgjnwN2EU.

35 Informational video, Tyler DeWhitt, "Electron Capture," Socratic.org, January 5, 2012, https://www.youtube.com/watch?v=sg_XoUDsP08.

36 See "Interactive Chart of the Nuclides," National Nuclear Data Center, Brookhaven National Laboratory, https://www.nndc.bnl.gov/nudat2/ .

37 K. P. Sinha, "A Theoretical Model for Low-Energy Nuclear Reactions," *Infinite Energy*, 29 (January/February 2000): 54–57.

38 Edmund Storms, "The Explanation of Low Energy Nuclear Reactions," Infinite Energy Press (2014).

39 Alan Widom and Lewis Larsen, "Ultra Low Momentum Neutron Catalyzed Nuclear Reactions on Metallic Hydride Surfaces," *Condensed Matter* (May 2, 2005): 1-4.

40 Colin David Scarfe, "The Gamma Radiation from the Bombardment of Heavy Ice with Low Energy Protons," The University of British Columbia (October 1961).

41 A. Machiels and T. O. Passell, "Development of Energy Production Systems from Heat Produced in Deuterated Metals: Volume 2," Electric Power Research Institute, TR-107843-V2 (November 1999).

42 Setauo Ichimaru, "Radiative Proton-Capture Nuclear Processes in Metallic Hydrogen," Physics of Plasmas, 8(10), (October 2001): 4284–91.

43 Glenn F. Knoll, "Radiation Detection and Measurement," John-Wiley & Sons, Inc. (2000).

44 See https://www.nist.gov/pml/xcom-photon-cross-sections-database.

45 D. J. Osias, "Status of Nuclear Flight System Definition Studies - Case 237," B71 0218 (NASA Contractor Report 116601), Bellcomm, Inc. (February 9, 1971).

46 J. H. Hubbell and Stephen M. Seltzer, "Tables of X-Ray Mass Attenuation Coefficients and Mass Energy-Absorption Coefficients from 1 keV to 20 MeV for Elements $Z = 1$ to 92 and 48 Additional Substances of Domestic Interest," National Institute of Standards and Technology (January 1, 1995).

47 E. D. Arnold, "Handbook of Shielding Requirements and Radiation Characteristics of Isotope Power Sources for Terrestrial, Marine and Space Applications," ORNL-3576, Oak Ridge National Laboratory (1964).

48 "Design Definition and Safety Evaluation Study of a Compact ^{60}Co Heat Source in Space," AGN-8441, Aerojet-General Corporation (September 1969).

49 "Detailed Specification, Part 1, Performance/Design and Qualification Requirements for Engine, NERVA, 75K, Full Flow," Data Item No. C002-CP090290A-F1, Aerojet Nuclear Systems Company (September 8, 1970).

50 See pages 6-10 in the report, "Fusion Energy Sciences Roundtable on Quantum Information Science," May 1-2, 2018, at https://science.osti.gov/-/media/fes/pdf/workshop-reports/FES-QIS_report_final-2018-Sept14.pdf.

51 See https://www.euro-fusion.org/fusion/fusion-conditions/.

52 See http://www.ans.org/pubs/journals/fst/.

53 See https://www.iter.org/.

54 See Raffi Khatchadourian, "A Star in a Bottle," the New Yorker, March 3, 2014, https://www.newyorker.com/magazine/2014/03/03/a-star-in-a-bottle .

55 See Guy Norris, "Skunk Works Reveals Compact Fusion Reactor Details," Aviation Week and Space Technology, October 21, 2014, https://fusion4freedom.com/skunk-works-reveals-compact-fusion-reactor-details/ .

56 Article in the *Washington Post*, (May 19, 2019): B1/B4.

57 See https://www.nichenergy.com.

58 Randolph R. Davis et al., "Electrolysis Apparatus and Electrodes and Electrode Material Therefor," US 6,248,221 B1 (June 19, 2001).

59 "Electro- and Thermo-transport of Hydrogen in Metals," by H. Wipf in *Hydrogen in Metals II. Application Oriented Properties*, G. Alefeid and J. Voikl (ed.), Springer-Verlag, Berlin (1978).

60 Ernest Charles Alcaraz et al., "Modular Gaseous Electrolysis Apparatus with Actively-Cooled Header Module, Co-Disposed Heat Exchanger Module and Gas Manifold Modules Therefor," US 10,465,302 B2 (November 5, 2019).

61 "The Gamma Spectrum Generator (GSG)," Joint Research Centre Institute for Transuranium Elements, Karlsruhe, Germany (See https://nucleonica.com/).

62 Monte S. Chawla et al., "Modular Cooling Chamber for Manifold of Gaseous Electrolysis Apparatus with Helium Permeable Element Therefor," US 10,480,084 B1 (November 19, 2019).

63 S. Focardi et al., "Energy Generation and Generator by Means of Anharmonic Stimulated Fusion," WO 95/20816 (August 3, 1995).

64 See Hyperion system design by Defkalion Green Technologies, S.A., Athens, Greece (https://www.nextbigfuture.com/2011/11/praxen-defkalion-reveals-technical.html).

---

65 A. Rossi, "Method and Apparatus for Carrying Out Nickel and Hydrogen Exothermal Reactions," US2011/0005506A1 (January 13, 2011).

66 Gyorgy Egely, "Method for Production of Renewable Heat Energy," US2014/0098920A1 (April 10, 2014).

67 See "World Integrated Trade Solutions (WTS), The World Bank, https://wits.worldbank.org/.

68 Andrew J. W. Physick et al., "Novel Low Energy Hydrogen-Deuterium Isotope Breakthrough Separation Using a Trapdoor Zeolite," *Chemical Engineering Journal*, 288 (15 March 2016): 161–68.

69 Ming Liu et al., "Barely Porous Organic Cages for Hydrogen Isotope Separation," *Science*, 366 (issue 6465) (November 1, 2019): 613–20.

70 See "5 Fundamentals of a Renewable Energy Microgrid" and "4 Ways That Hybrid Electric Power Systems Can Reduce Facility Operating Cost," On-Line Training, Caterpillar Inc., https://www.cat.com/en_US/articles/support/electric-power-generation/power-systems-training.html.

APPENDIX

Modular Gaseous Electrolysis Apparatus

US010465302B2

(12) **United States Patent**
    Alcaraz et al.

(10) **Patent No.:** US 10,465,302 B2
(45) **Date of Patent:** Nov. 5, 2019

(54) **MODULAR GASEOUS ELECTROLYSIS APPARATUS WITH ACTIVELY-COOLED HEADER MODULE, CO-DISPOSED HEAT EXCHANGER MODULE AND GAS MANIFOLD MODULES THEREFOR**

(71) Applicant: **Marathon Systems, Inc.**, Fairfax, VA (US)

(72) Inventors: **Ernest Charles Alcaraz**, Vienna, VA (US); **Monte S. Chawla**, University Park, MD (US); **Randolph R. Davis**, Fairfax, VA (US); **Dorin A. Jannotta**, Jacksonville, FL (US); **Austin Lowrey, III**, Lancaster, PA (US); **Thomas F. McGraw**, New Bern, NC (US); **Frederick Sandel**, Fairfax Station, VA (US); **Donald J. Waltman**, Pasadena, MD (US)

(73) Assignee: **Marathon Systems, Inc.**, Fairfax, VA (US)

(*) Notice: Subject to any disclaimer, the term of this patent is extended or adjusted under 35 U.S.C. 154(b) by 0 days.

(21) Appl. No.: **15/438,768**

(22) Filed: **Feb. 22, 2017**

(65) **Prior Publication Data**
    US 2018/0087165 A1   Mar. 29, 2018

Related U.S. Application Data

(63) Continuation-in-part of application No. 14/815,935, filed on Jul. 31, 2015, now abandoned.
    (Continued)

(51) **Int. Cl.**
    C25B 15/02   (2006.01)
    C25B 9/12   (2006.01)
    (Continued)

(52) **U.S. Cl.**
    CPC *C25B 15/02* (2013.01); *C25B 9/00* (2013.01); *C25B 9/12* (2013.01); *C25B 15/08* (2013.01); *G21B 3/00* (2013.01); *G21C 3/38* (2013.01)

(58) **Field of Classification Search**
    CPC C25B 1/02–1/12
    (Continued)

(56) **References Cited**

U.S. PATENT DOCUMENTS

3,290,522 A	12/1966	Ginell	
3,409,820 A	11/1968	Burke	
(Continued)			

FOREIGN PATENT DOCUMENTS

CN	101395677 B	7/2012
DE	102006007773 A1	9/2007
(Continued)		

OTHER PUBLICATIONS

"Modular design", Wikipedia, available at https://en.wikipedia.org/wiki/Modular_design, accessed on Jan. 10, 2019 (Year: 2019).*
(Continued)

Primary Examiner — Harry D Wilkins, III
(74) *Attorney, Agent, or Firm* — ATFirm PLLC; Ralph P. Albrecht

(57) **ABSTRACT**

An improved, gaseous electrolysis apparatus can include a cooled header for electric connections or couplings, an exemplary co-disposed, coaxial heat exchanger around the reaction chamber to extract heat from the reaction chamber and exemplary rugged gas source and collection manifold(s) to support fixed and/or mobile applications in an embodiment. The system can include a heated anode and co-disposed cylindrical cathode within the reaction chamber and an improved electronic control circuit in an embodiment.

20 Claims, 15 Drawing Sheets

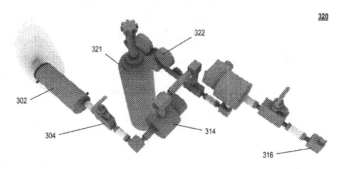

126

Related U.S. Application Data

(60) Provisional application No. 61/999,824, filed on Aug. 7, 2014.

(51) **Int. Cl.**

C25B 15/08	(2006.01)
C25B 9/00	(2006.01)
G21B 3/00	(2006.01)
G21C 3/38	(2006.01)

(58) **Field of Classification Search**

USPC .... 204/272; 123/3; 174/151–152 GM
See application file for complete search history.

(56) **References Cited**

U.S. PATENT DOCUMENTS

3,755,128 A	8/1973	Herwig	
4,117,254 A	* 9/1978	Richter	B03C 3/70
			174/15.3
4,174,145 A	11/1979	Oeschger	
4,265,721 A	* 5/1981	Hackmyer	C25B 1/04
			205/340
4,593,758 A	6/1986	Kyle	
4,835,433 A	5/1989	Brown	
4,867,228 A	9/1989	Novelli et al.	
5,075,160 A	12/1991	Stinton et al.	
5,230,729 A	7/1993	Mccandlish	
5,273,203 A	12/1993	Webster	
5,273,635 A	12/1993	Gernert et al.	
5,318,675 A	6/1994	Patterson	
5,352,269 A	10/1994	Mccandlish	
5,366,712 A	11/1994	Violante et al.	
5,395,422 A	3/1995	Schulz et al.	
5,411,654 A	5/1995	Ahern et al.	
5,429,725 A	7/1995	Thorpe et al.	
5,472,614 A	12/1995	Rossi	
5,569,561 A	10/1996	Exnar et al.	
5,674,632 A	10/1997	Ahern et al.	
5,770,036 A	6/1998	Ahern et al.	
6,051,110 A	4/2000	Dell'Orfano et al.	
6,248,221 B1	6/2001	Davis et al.	
6,620,994 B2	9/2003	Rossi	
7,244,887 B2	7/2007	Miley	
7,579,117 B1	8/2009	Beard	
7,767,066 B2	8/2010	May	
8,051,637 B2	11/2011	Gaudencio Aquino	
8,227,020 B1	7/2012	Miley	
8,264,382 B2	9/2012	Rigolle et al.	
8,303,865 B1	11/2012	Cravens	
8,419,919 B1	4/2013	Boss et al.	
8,603,405 B2	12/2013	Miley	
8,636,881 B2	1/2014	May	
8,652,319 B2	2/2014	Kothe	
8,679,326 B2	3/2014	Vinci et al.	
2002/0046762 A1	4/2002	Rossi	
2004/0084326 A1	* 5/2004	Weinberg	C25B 1/02
			205/628
2005/0105664 A1	5/2005	Chubb	
2005/0120715 A1	6/2005	Labrador	
2005/0199747 A1	9/2005	Roarty	
2007/0170051 A1	7/2007	Schlaikjer	
2007/0280398 A1	12/2007	Dardik	
2008/0017504 A1	* 1/2008	Liu	C25C 3/085
			204/228.3
2008/0159461 A1	7/2008	Chubb	
2008/0205572 A1	8/2008	Chubb	
2010/0123022 A1	5/2010	Roarty	
2010/0259422 A1	10/2010	Rigolle et al.	
2010/0275859 A1	* 11/2010	Klotz	C25B 1/06
			123/3
2010/0300945 A1	12/2010	Vinci et al.	
2011/0005506 A1	1/2011	Rossi	
2011/0044419 A1	2/2011	Cook	
2011/0155566 A1	* 6/2011	Martinez Cao	C25B 1/04
			204/269
2011/0247929 A1	* 10/2011	Nagai	C02F 1/46109
			204/290.15
2011/0253527 A1	* 10/2011	Hui	C02F 1/46104
			204/267
2012/0008728 A1	1/2012	Fleming	
2013/0044847 A1	2/2013	Steinberg	
2013/0233718 A1	9/2013	Roarty	
2013/0243143 A1	9/2013	Mastromatteo et al.	
2014/0099252 A1	4/2014	Chason et al.	
2014/0202877 A1	* 7/2014	Greenbaum	C25B 11/02
			205/630
2014/0326711 A1	11/2014	Rossi	
2014/0332087 A1	11/2014	Godes et al.	
2015/0000252 A1	1/2015	Moore et al.	
2015/0211131 A1	* 7/2015	Jacobs	C25B 15/02
			204/236
2017/0088958 A1	* 3/2017	Koeneman	C25B 11/02

FOREIGN PATENT DOCUMENTS

DE	102006007773 B4	2/2010	
EP	0568118 A2	11/1993	
GB	2231195 A	11/1990	
JP	S6270203 A	3/1987	
JP	63-282286 A	* 11/1988	C25B 9/00
WO	WO1990010935 A1	9/1990	
WO	WO1990014669 A1	11/1990	
WO	WO 1995/020816	8/1995	
WO	WO1995020816 A1	8/1995	
WO	WO 2004044924 A1	5/2004	
WO	WO 2007096120 A2	8/2007	
WO	WO 2007130156 A2	11/2007	
WO	WO 2013170244 A2	11/2013	
WO	WO 2013170244 A3	1/2014	

OTHER PUBLICATIONS

Egan, Implementing a Successful Modular Design—PTC's Approach, 7th Workshop on Product Structuring—Product Platform Development, Chalmers Univeristy of Technolog, Goteborg, Germany, Mar. 2004, pp. 49-58 (Year: 2004).*

Martin Fleischmann and Stanley Pons, "Electrochemically Induced Nuclear Fusion of Deuterium," submitted to the Journal of Electroanalytical Chemistry, Mar. 11, 1989.

Thomas F. McGraw and Randolph R. Davis, "Critical Factors in Transitioning from Fuel Cell to Cold Fusion Technology," 33rd Intersociety Engineering Conference on Energy Conversion (IECEC-98-1271), Colorado Springs, CO, Aug. 2-6, 1998.

David Nagel et al., briefing charts from "Perspectives on Low Energy Nuclear Reactions Workshop/Short Course," NUCAT Energy, LLC, held at Hyatt Regency Crystal City, VA, Oct. 3-4, 2011.

Bob Sterling, "The Nuclear Reactor in Your Basement," NASA Newsletter on Global Climate Change, Feb. 12, 2013.

R.G. Bosisio et al., "The Large Volume Microwave Plasma Generator (LMP™): A New Tool for Research and Industrial Processing," Journal of Microwave Power, 7(4), 1972.

Yu.A.Lebedev, "Microwave Discharges: Generation and Diagnostics," 25th Summer School and International Symposium on the Physics of Ionized Gases (SPIG 2010), Journal of Physics: Conference Series 257 (2010) 012016.

K.P. Sinha, "A Theoretical Model for Low-Energy Nuclear Reactions in a Solid Matrix," Infinite Energy Magazine, Issue 29, 1999.

F. A. Leavitt et al., "Use Application and Testing of Hi-Z Thermoelectric Modules," Hi-Z Technology, Inc., 2007.

David S. Alexander, "Advanced Energetics for Aeronautical Applications," vol. I, NASA/CR-2003-212169, Feb. 2003, and vol. II, NASA/CR-2005-213749, Apr. 2005.

Marty K. Bradley and Christopher K. Droney, "Subsonic Ultra Green Aircraft Research—Phase II: N+4 Advanced Concept Development," NASA/CR-2012-217556, May 2012.

"Energy Loss and Range of Electrons and Pinions," National Bureau of Standards Circular 577, 1956. No month and/or year given.

(56)   **References Cited**

OTHER PUBLICATIONS

A.G. Lipson, DM Sakov, V.B. Kalinin, E.I. Saunin and B.V. Derjaguin, "Observation of Neutrons and Tritium in Kd2PO4 Single Crystals Upon the Ferroelectric Phase Transition," Fourth International Conference on Cold Fusion, Dec. 6-9, 1993.

B. Danapani and M. Fleischman. "Electrotytic Separation Factors on Palladium," Journal of Electrochemistry, 39, pp. 323-332, 1972. No month and/or year given.

Brian D. Andresen, Richard Whipple, Armando Alcaraz, Jeffrey S. Haas, and Patrick M. Grant, "Potentially Explosive Organic Reaction Mechanisms in Pd/D2O Electrochemical Cells," Chemical Health & Safety, 1(3), pp. 44-47, Oct./Nov. 1994.

D.L. Donohue and Milica Petek, "Isotopic Measurements of Palladium Metal Containing Protium and Deuterium by Glow Discharge Mass Spectrometry," Analy. Chem. 63, p. 740-744, 1991. No month and/or year given.

D.P. Stinton, T.M. Besmann, S. Shanmugham, A. Bleier, E. Lara-Curzio, "Development of Osidation/Corrosion-Resistant Composite Materials and Interfaces," Fossile Energy Program Annual Progress Report for Apr. 1994-Mar. 1995, ORNL-6874, pp. 21-33, 1995.

D.W. Mo, Q.S. Cai, L.M. Wang, S.Z. Wang, "The Evidence of Nuclear Transmutation Phenomeno in Pd-H System Using NAA (Neutron Activation Analysis)," ICCR-7, Vancouver, pp. 259-263. No month and/or year given.

Debra R. Rolison and William E. O'Grady. "Observation of Elemental Anomalies at the surface of Palladium after Electrochemical Loading of Deuterium or Hydrogen," Analytical Chemistry, 63(17), pp. 1697-1702, Sep. 1, 1991, D.W. Mo, Q.S. Cai, L.M. Wang, S.Z.

Eiichi Yamaguchi and Hiroshi Sugiura. Excess Heat and Nuclear Products from Pd:D/Au Heterostructures by the 'In-vacno' Method, ICCF-7, Vancouver, pp. 420-424. No month and/or year given.

Eiichi Yamaguchi and Hiroshi Sugiura. Excess Heat and Nuclear Products from Pd:D/Au Heterostructures by the 'In-vacuo' Method, ICCF-7, Vancouver, pp. 420-424. No month and/or year given.

Ex parte Dash. No. 92-3536, U.S. Patent and Trademark Office Board of Patent Appeals and Interferences, 27 USPQ2d, pp. 1481-1492. No month and/or year given.

F. Piantelli, S. Focardi and R. Habel, "Energy Generation and Generator by Means of Anharmonic Stimulated Fusion," International Patent reproduced in Infinite Energy. Sep.-Oct. 1995, pp. 24-31. No month and/or year given.

G. Mengoli et al. "Calorimetry Close to the Boiling Temperature of the D2O/Pd Electrolytic System," Journal of Electroanalytical Chemistry, 444, pp. 155-167, 1998. No month and/or year given.

Heinrich Hora, George H. Miley, Jak C. Kelly, and Y. Narne, "Nuclear Shell Magic Numbers Agree With Measured Transmutation by Low-Energy Reactions," ICCF-7, Vancouver, pp. 147-151. No month and/or year given.

Hydrogen in Metals I: Basic Properties, G. Alefeld and J. Volkl (ed.), Chapters ;8 and 12, Springer-Verlag, Berlin, 1978. No month and/or year given.

Hydrogen in Metals II: Application-Oriented Properties, G. Alefeld and J. Volkl (ed.), Chapter 7, Springer-Verlag, Berlin, 1978.

Jellinek, "Theoretical Dynamical Studies of Metal Clusters and Cluster-Ligand Systems," Metal-Ligand Interactions: Structure and Reactivity, N. Russo (ed.), Kluwer-Dordrecht, 1995. No month and/or year given.

Ray E. Kidder, "Energy Transfer Between Charged Particles by Coulomb Collosions," URL-5213, University of California Radiation Laboratory, Livermore, California, May 12, 1958.

Robert J. LeRoy, Steven G. Chapman, and Frederick R. W. McCourt, "Accurate Thermodynamic properties of the Six Isotopomers of Diatomic Hydrogen," The Journal of Physical Chemistry, 94(2), pp. 923-929, 1990. No month and/or year given.

S. Focardi, R. Habel, and F. Piantelli, Anomalous Heat Production in Ni—H Systems, Il Nuovo Cimento, 107A(1), pp. 163-167, Feb. 1994.

S. Srinivasan, "Fuel Cells for Extraterrestrial and Terrestrial Applications," Journal of the Electrochemical Society, 136(2), 41-48C, Feb. 1989.

S. Ueda, K. Yasuda and A. Takahashi, "Study of Excess Heat and Nuclear Products with Closed Electrolysis System and Quadrupole Mass Specirometer," ICCF-7, Vancouver, pp. 398-402, No month and/or year given.

S.W. Stafford and R.B. McLellan, "The Solubility of Hydrogen in Nickel and Cobalt," Acta Metallurgica. 22, pp. 1463-1468, 1974. No month and/or year given.

Talbot A. Chubb and Scott R. Clubb, "Cold Fusion as an interaction Between Ion Band States," Fusion Technology, 20, pp. 93-99, Aug. 1991.

Y. Arata and Y. Zhang, "Helium (42He, 32He) within Deuterated Pd-Black," Proceeding of the Japanese Academy, vol. 73(B), No. 1, pp. 1-6 (1997). No month and/or year given.

Y. Oya, H. Ogawa, M. Aida, K. Iinuma and M. Okamoto, "Material Conditions to Replicate the Generation of Excess Energy and the Emission of Excess Neutrons," ICCF-7, Vancouver, pp. 285-291. No month and/or year given.

U.S. Appl. No. 61/999,824, filed Aug. 7, 2014.

Martin Fleischmann and Stanley Pons, "Electrochemically Induced Nuclear Fusion of Deuterium," Journal of Electroanalytical Chemistry, 261, pp. 301-, 1989.

M.C.H. McKubre et al., "Excess Power Observations in Electrochemical Studies of the D/Pd System: The Influence of Loading," 3rd International Conference on Cold Fusion, Frontiers of Cold Fusion, Nagoya, Japan, 1992, Universal Academy Press, Inc., Tokyo, Japan.

K. Kunimatsu et al., "Deuterium Loading Ratio and Excess Heat Generation during Electrolysis of Heavy Water by Palladium Cathode in a Closed Cell using a Partially Immersed Fuel Cell Anode," 3rd International Conference on Cold Fusion, Frontiers of Cold Fusion, Nagoya, Japan, 1992, Universal Academy Press, Inc., Tokyo, Japan.

V. Violante et al., "RF Detection and Anomalous Heat Production during Electrochemical Loading of Deuterium in Palladium," Energia, Ambiente e Innovaziene, 2-3, pp. 63-77, 2014.

M. Miles et al., "Correlation of Excess Power and Helium Production during D2O and H2O Electrolysis using Palladium Cathodes," Journal of Electroanalytical Chemistry, 346, p. 99-, 1993.

M.C.H. McKubre et al., "The Emergence of a Coherent Explanation for Anomalies Observed in D/Pd and H/Pd System: Evidence for 4He and 3He Production," 8th International Conference on Cold Fusion, 2000, Lerici (La Spezia), Italy: Italian Physical Society, Bologna, Italy.

P.A. Boss et al., "Investigation of Nano-Nuclear Reactions in Condensed Matter," U.S. Department of Defense, 2016.

T. McGraw and R. Davis, "Critical Factors in Transitioning from Fuel Cell to Cold Fusion Technology," 33rd Intersociety Engineering Conference on Energy Conversion (IECEC-98-1271), Colorado Springs, CO, Aug. 2-6, 1998.

M. Chawla and R. Davis, "Key Issues Related to Industrialization of LENR-Based Space Propulsion," Space Technology & Applications Industrial Forum, Apr. 2016 (draft).

J. J. Thomson, "On the Electrolysis of Gases," Proceedings of the Royal Society, 58, No. 350, pp. 244-257, Jun. 1885.

F. A. Maxfield f R. R. Benedict, "Theory of Gaseous Conduction and Electronics," pp. 270-274 and 293-294, McGraw-Hill, 1941.

S. C. Brown, "Introduction to Electrical Discharges in Gases," pp. 188-190, John Wiley & Sons, 1966.

A. V. Phelps, "Cross Sections and Swarm Coefficients for H+, H2+, H3+, H, H2 and H– in H2 for Energies from 01. eV to 10 keV," Journal of Physical Chemistry, Reference Data, 19, No. 3, 1990.

R.G. Bosisio et al., "The Large Volume Microwave Plasma Generator (LMPTM): A New Tool for Research and Industrial Processing," Journal of Microwave Power, 7(4), 1972.

Davis, Randolph R., "Technical Background for Mk12.31 Program," NEPS-TN-004, New Energy Power Systems LLC, Fairfax, VA 22038, dated Mar. 29, 2018, published/submitted Apr. 5, 2018.

* cited by examiner

FIG. 1A

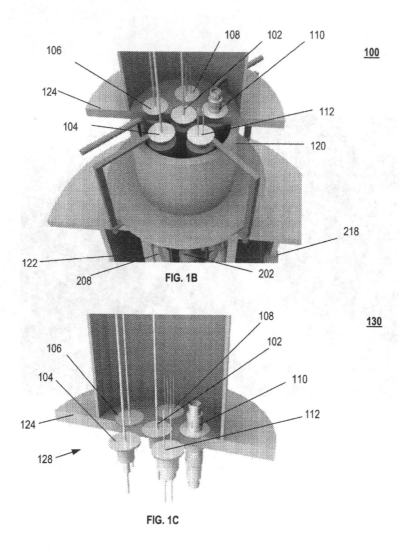

FIG. 1B

FIG. 1C

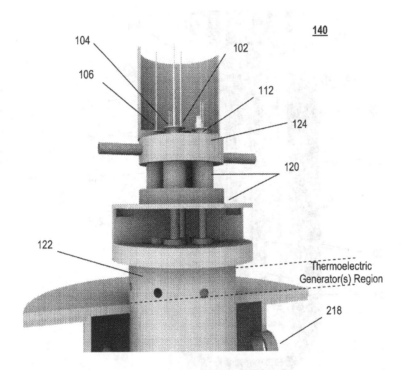

FIG. 1D

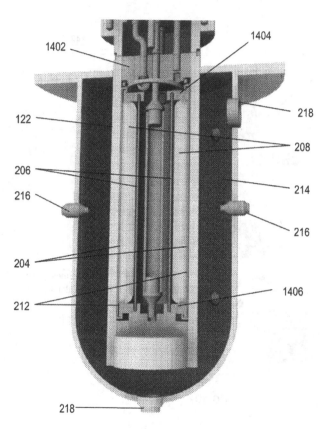

FIG. 2A

230

216

212

1406

208

206

204

202

FIG. 2B

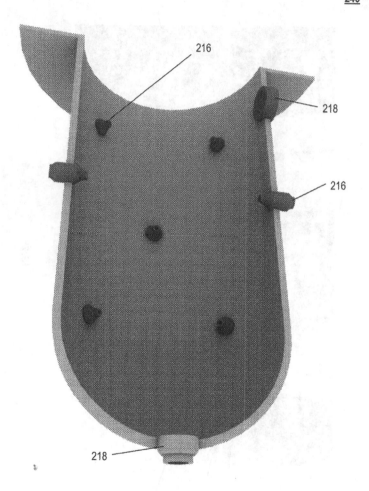

240

216

218

216

218

FIG. 2C

134

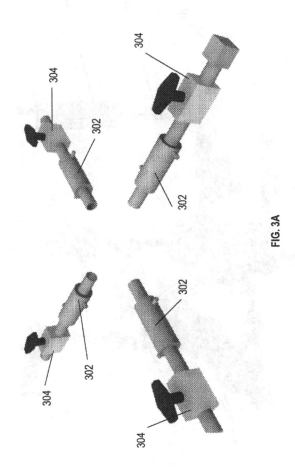

300

304

304

302

302

302

302

304

304

FIG. 3A

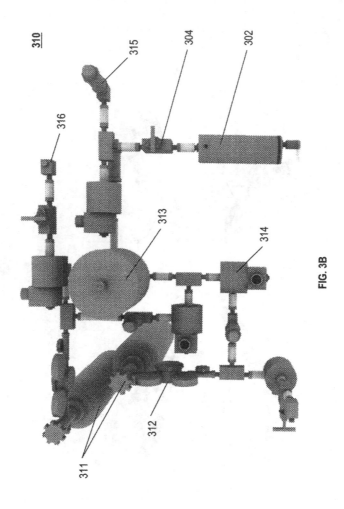

310

315

304

302

316

313

314

311

312

FIG. 3B

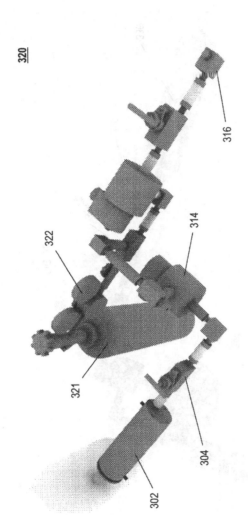

320

322

321

302

304

314

316

FIG. 3C

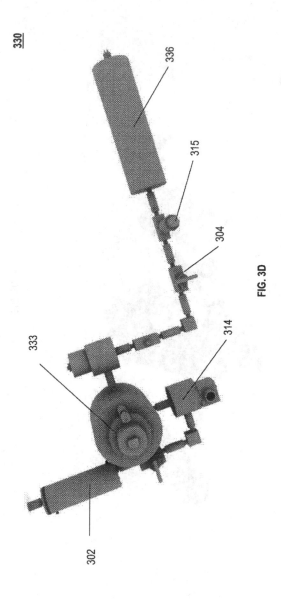

330

336

315

304

314

333

302

FIG. 3D

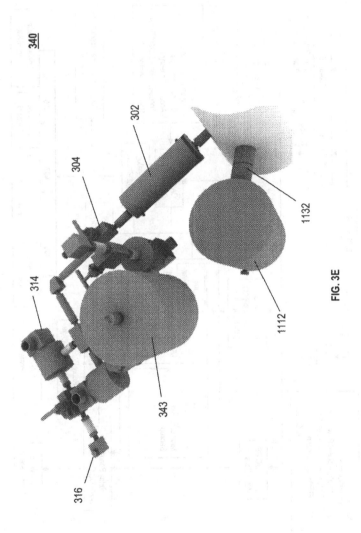

FIG. 3E

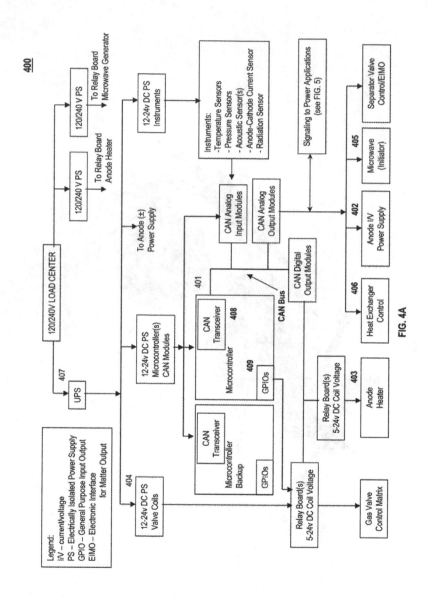

FIG. 4A

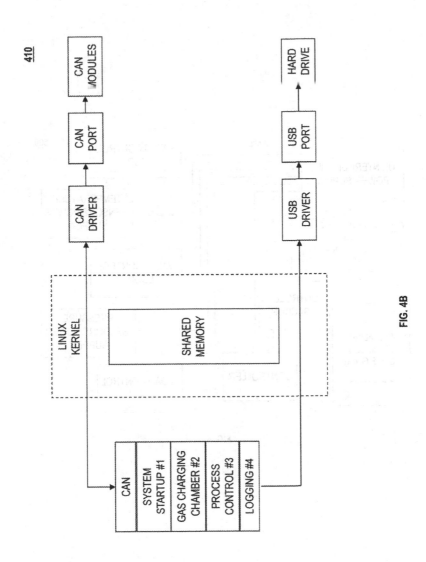

FIG. 4B

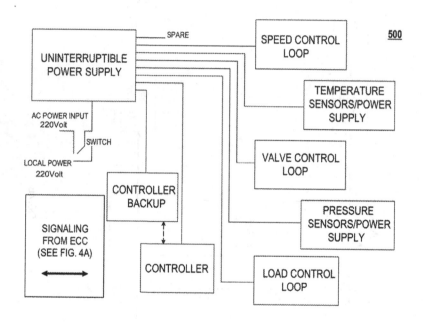

FIG. 5

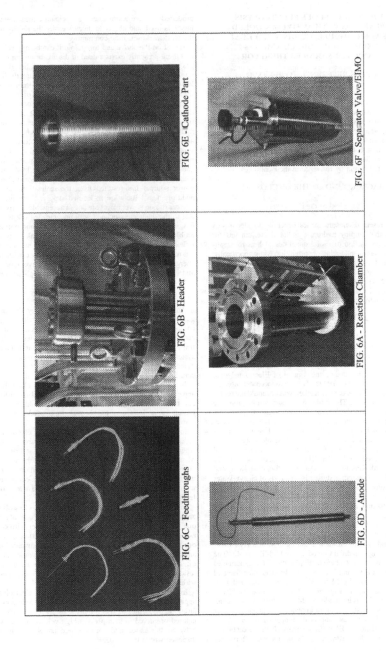

FIG. 6E - Cathode Part

FIG. 6F - Separator Valve/EIMO

FIG. 6B - Header

FIG. 6A - Reaction Chamber

FIG. 6C - Feedthroughs

FIG. 6D - Anode

1   2

MODULAR GASEOUS ELECTROLYSIS APPARATUS WITH ACTIVELY-COOLED HEADER MODULE, CO-DISPOSED HEAT EXCHANGER MODULE AND GAS MANIFOLD MODULES THEREFOR

CROSS-REFERENCE TO RELATED APPLICATIONS

This Application claims benefit under 35 U.S.C. § 120 and is a continuation-in-part of U.S. NonProvisional patent application Ser. No. 14/815,935 filed Jul. 31, 2015, presently pending, which claims benefit under 35 U.S.C. § 119(e) of U.S. Provisional Patent Application No. 61/999,824 filed Aug. 7, 2014, both of which are of common assignee to the present application, the contents of both of which are incorporated herein by reference in their entireties.

BACKGROUND OF THE DISCLOSURE

1. Field of the Disclosure

The present disclosure relates generally to electrolysis systems that employ hydrogen and/or deuterium gas for practical production of useful quantities of heat for application to many challenging opportunities. More specifically, the present disclosure relates to improved materials, structures and methods for improving low energy nuclear reaction (LENR) systems. This invention specifically addresses the technical challenges of significant energy production in an industrially viable configuration.

2. Description of the Related Art

In 1989, Martin Fleischmann and Stanley Pons announced the discovery of anomalous heat production in electrolytic cells of deuterium oxide and palladium (see, e.g., ref "Electrochemically Induced Nuclear Fusion of Deuterium," submitted to the Journal of Electroanalytical Chemistry, Mar. 11, 1989). The announcement became extremely controversial when other workers could not reproduce their results. The difficulty in part results from the problem of properly loading deuterium into the cathode. Over the intervening years, however, many other workers have reported positive results. Examples of evidence showing excess heat is obtainable in liquid electrolysis cells include: "Excess Power Observations in Electrochemical Studies of the D/Pd System: The Influence of Loading," by M. C. H. McKubre et al.; "Deuterium Loading Ratio and Excess Heat Generation during Electrolysis of Heavy Water by Palladium Cathode in a Closed Cell using a Partially Immersed Fuel Cell Anode," by K. Kunimatsu et al.; and "RF Detection and Anomalous Heat Production during Electrochemical Loading of Deuterium in Palladium," by V. Violante et al. Examples of evidence showing fusion by-products are obtainable include: "Correlation of Excess Power and Helium Production during D$_2$O and H$_2$O Electrolysis using Palladium Cathodes," by M. Miles et al.; "The Emergence of a Coherent Explanation for Anomalies Observed in D/Pd and H/Pd System: Evidence for ^{4}He and ^{3}He Production," by M. C. H. Mckubre et al.; and U.S. Department of Defense report, "Investigation of Nano-Nuclear Reactions in Condensed Matter," by P. A. Boss et al. Today, there exists a body of knowledge contained in over a thousand books and papers documenting the advances made in the field of LENR (Low Energy Nuclear Reactions). Operability is demonstrated by the numbers of efforts that produced positive results, accurate scientific measurement methods used, and ability to discuss easily the results with the worldwide scientific community.

Based on the above and foregoing, it can be appreciated that there presently exists a need in the art for an improved gas electrolysis apparatus that overcomes difficulties in up-scaling of liquid LENR systems and that directly links laboratory experimental apparatuses with a commercially useful device for producing energy at levels of practical interest. The present disclosure was also motivated by a desire to overcome issues in future development of conventional energy sources. Conventional energy systems produce hydrocarbon emissions believed to cause climate changes and that threaten human, animal and plant existence. Conventional energy systems fail to provide modular, reliable, rugged and self-contained power sources providing heat and/or electricity in applications ranging from fixed power sources to mobile vehicles. A sense of urgency for such new power sources has resulted from the realization that, in addition, fossil fuels are in limited supply. Conventional nuclear fission power plants are not an acceptable alternative due to the dangers associated with uncontrolled releases of fission products, the enormous environmental and political problems associated with nuclear waste disposal, costs and inflexible designs. In particular, there is widespread need for innovative energy sources that do not rely on established energy infrastructures. Furthermore, cost-effective, modular and scalable designs are needed to increase applicability.

SUMMARY OF THE INVENTION

An example embodiment of the invention may include: a gaseous electrolysis apparatus, which can include: a) a cooled header with at least one electrical connector or coupling; b) a heat exchanger configured to remove heat from a surface of a reaction chamber; c) a gas handling system mechanically coupled to the reaction chamber; and d) an electronic control circuit (ECC) electrically coupled or connected to the header and gas handling system.

In one example embodiment, the gaseous electrolysis apparatus can include where the header can include at least one of: a cooling apparatus; a cooling manifold or water jacket; or at least one feedthrough to a header cooling manifold proximate to the header.

In one example embodiment, the gaseous electrolysis apparatus can include the at least one feedthrough, and where the at least one feedthrough can include at least one of: a pressure side oriented towards the inside of the reaction chamber; is welded into a thermal plate; wherein one end of the feedthrough extends beyond the header for connection with the electronic control circuit; or wherein the at least one feedthrough comprises a threaded coupling.

In one example embodiment, the gaseous electrolysis apparatus can include where the header can include: at least one anode connection; at least one anode heater wire connection; at least one microwave antenna connection; at least one thermal sensor connection; at least one microwave loop antenna; and at least one insulator configured to at least one of: electrically isolate; minimize the volume where gas resides; or provide mechanical support for components within the reaction chamber.

In one example embodiment, the gaseous electrolysis apparatus can further include a modular, removable anode; an anode where the edges of the anode facing the cathode are tapered or curved to help prevent high voltage breakdown between the anode and cathode; and an electric heater disposed within the anode.

3

In one example embodiment, the gaseous electrolysis apparatus can further include a modular, removable, hollow-shaped, cylindrical cathode with a central cavity configured to receive the anode; a cathode encased by an outer metal supporting sleeve; a cathode bounded at its base and top with insulator endcaps; and a cathode wherein the edges of the reaction material part of the cathode facing the anode are tapered or curved to help prevent high voltage breakdown.

In one example embodiment, the gaseous electrolysis apparatus can include where the heat exchanger is modular and can include: a relatively low volume flash boiler configured to provide a mist of water or other coolant to the outer surface of the reaction chamber; is co-disposed around the reaction chamber; a plurality of spray nozzles to cool at least one portion of the reaction chamber; at least one steam pressure port; and at least one thruster port configured to provide pressure output.

In one example embodiment, the gaseous electrolysis apparatus can include where the gas handling system can include: four (4) separate gas manifolds that control gas flow while minimizing gas volume external to the reaction chamber, where the four separate gas manifolds can include: a hydrogen/deuterium gas supply manifold; an inert carrier gas manifold; a reaction gas product collection manifold; and a gas measurement and evacuation manifold.

In one example embodiment, the gas electrolysis apparatus can include where the gas manifold can include: a cooling chamber or water jacket to provide cooling for gas tubing and pipes connected or coupled to the reaction chamber; at least one gas compatible valve, at least one pressure sensor and at least one temperature sensor connected to, or coupled to the electronic control circuit/subsystem; at least one tank or at least one container whose known volume enables small quantities of gas to be determined by calculating pressure, temperature and volume before gas is transferred into or out of the reaction chamber; and at least one purge port which can be used to evacuate gases manually from containments of the gas manifolds.

In one example embodiment, the gas electrolysis apparatus can include a reaction gas product collection manifold wherein the reaction gas product collection manifold further comprises: a container configured to temporarily store reactant gas and periodically permit extraction from the container; and a subsystem comprising an acoustic sensor or other type of electronic interface to facilitate estimation of quantities of reaction product gas.

In one example embodiment, the gaseous electrolysis apparatus according to claim 1, wherein the electronic control circuit (ECC) further comprises: a special-purpose computer and display monitor, the special-purpose computer comprising at least one of a microprocessor or a microcontroller, and comprising control software; special-purpose anode-to-cathode voltage/current supply; at least one anode heater supply; at least one microwave starter or initiator electronics; and at least one heat exchanger electronics.

The present application provides an improved apparatus or system utilizing gaseous electrolysis techniques according to an exemplary embodiment, including: (1) an exemplary cooled header for electric connections; (2) an exemplary co-disposed, coaxial heat exchanger/thruster around an exemplary reaction chamber to extract heat from, or perform testing of, the reaction chamber; and (3) exemplary rugged, modular gas source and gas collection manifolds. In addition, the application provides an exemplary modular anode; an exemplary modular, cylindrical cathode co-disposed around the anode; and, an improved modular electronic control circuit for control of apparatus operation.

4

One aspect of an exemplary embodiment of the present disclosure is to provide a modular electrolysis apparatus. According to one aspect of the present disclosure, the modular apparatus can enable a number of system houses to manufacture, install, repair and otherwise support industrialization objectives.

An aspect of an exemplary embodiment of the present disclosure is to provide an electrolysis apparatus which can utilize a new header design for electrical power, device control and instrumentation connections or couplings. One aspect of the disclosure, in an exemplary embodiment, is that the header advantageously enables the apparatus to be serviced when internal components are expended or may require maintenance. According to another aspect of an exemplary embodiment of the disclosure, the header can include components that are adapted to be cooled to advantageously enable the apparatus to be operated for longer times between maintenance periods.

Another aspect of an exemplary embodiment of the present disclosure is to provide an exemplary heat exchanger module to extract heat from the electrolysis system. According to another aspect of an exemplary embodiment of the present disclosure, the heat exchanger module can facilitate and control a thermal gradient through the cathode module and the wall of the reaction chamber module.

Yet another aspect of an exemplary embodiment of the present disclosure is to provide gas source and collection manifolds for electrolysis apparatuses. According to another aspect of an exemplary embodiment of the present disclosure, these gas handling manifolds can be designed and engineered to be rugged and compact to support fixed and/or mobile operation.

According to another aspect of an exemplary embodiment of the present disclosure, the inventive electrolysis apparatus can apply to an exemplary gas electrolysis apparatus having a cylindrical configuration which can utilize the above-mentioned improved exemplary header, co-disposed cylindrical heat exchanger and application specific gas handling manifolds. It should be mentioned that the exemplary cylindrical configuration of the electrolysis apparatus, according to an exemplary embodiment, can take advantage of radial and variable electric fields, gas pressure and thermal diffusion gradients, promoting gas transport into and through reaction material in the cathode.

Another aspect of an exemplary embodiment of the present disclosure is to provide a modular/removable anode within the reaction chamber that contains an internal electric heater and that can, by thermal radiation and diffusion, raise the temperature of the reaction material in the cathode.

Another aspect of an exemplary embodiment of the present disclosure is to provide a modular cathode that can be cylindrical and co-disposed around the anode within the reaction chamber, and which can include an exemplary consolidated metal powder and be encased with a supporting sleeve according to an exemplary embodiment. These improvements can provide further advantages in addition to the co-disposed cathode and consolidated cathode reaction material comprising nanocrystalline particles described earlier in U.S. Pat. No. 6,248,221 B1, the content of which is incorporated herein by reference in its entirety.

These and other aspects, features and advantages according to an exemplary embodiment of the present disclosure are provided by the exemplary header module that can be a modular unit and can include one or more exemplary high-temperature electric feedthroughs, one or more exemplary high-pressure gasket or gaskets, an exemplary cooling manifold or water jacket, exemplary electric wiring, exemplary connectors and exemplary supporting ceramic insulators.

The feedthroughs can be constructed with appropriately sized conductors and rugged insulating materials as illustrated in the accompanying drawing figures. According to an aspect of the disclosure, the exemplary electric feedthroughs and gasket(s) can enable high-pressure operation, as well as a vacuum startup environment, within the exemplary reaction chamber module to be maintained and isolated from the local external environment during operation. According to another exemplary aspect of the present disclosure, the exemplary electric feedthroughs enable internal components of the electrolysis system to be connected or coupled to external components of its electronic control circuit module. According to another aspect of the present disclosure, the cooling manifold or water jacket can maintain sufficiently low temperature during operation of the electrolysis apparatus for the feedthroughs and gasket(s) to maintain mechanical integrity over long periods of operation, in an exemplary embodiment. According to yet another aspect of the present disclosure, the exemplary electric wiring, connectors or couplers and ceramic insulators can be designed so that they connect to, or couple and provide mechanical support to the apparatus's high-voltage, high-temperature anode module, thermal sensor and any other necessary electric connections or couplings within the reaction chamber module, such as to an initiator/microwave antenna. In an exemplary embodiment, the exemplary header module can be able to be installed (closed) and later removed (opened) for maintenance with exemplary anode module, initiator/ microwave antenna, and thermal sensor attached.

These and other aspects, features and advantages according to an exemplary embodiment of the present disclosure are provided by an exemplary heat exchanger that can surround and can be coaxial to the reaction chamber of the electrolysis system in one embodiment. According to one exemplary aspect of the present disclosure, the heat exchanger can be a relatively low volume flash boiler engineered to provide a mist of water or other appropriate coolants to the surface of the reaction chamber. Advantageously, the resulting comparatively low temperature of the surface of the exemplary reaction chamber can produce a thermal gradient to facilitate thermal diffusion of gases through the cathode over long periods of operation, and, therefore, can be important to the heat producing process within the reaction material. According to another aspect of the disclosure, the exemplary heat exchanger can include spray nozzle(s), coolant pressure regulator(s) and/or a coolant pump that can be regulated and/or controlled to provide appropriate quantities of coolant to the surface of the reaction chamber. According to another aspect of the present disclosure, the exemplary heat exchanger can include one or more steam pressure port(s) and/or steam pressure regulator (s) to maintain a sufficiently high steam pressure for heating and power applications.

The exemplary electrolysis apparatus can preferably include a hydrogen/deuterium gas supply manifold, an inert carrier gas manifold, a reaction gas manifold and/or a gas measurement and evacuation manifold as part of an exemplary gas handling subsystem. The exemplary manifolds can each include individual cooling chambers, exemplary high-pressure-rated tubing and fittings, electrically and mechanically operated valves, measurement chambers and electrically-operated pressure and temperature sensors, according to an exemplary embodiment. According to one aspect of an exemplary embodiment of the present disclosure, the mani-

folds can be designed and engineered to be rugged, and occupy a minimum volume around the reaction chamber to support safety requirements. According to another aspect of an exemplary embodiment of the present disclosure, the gas handling manifolds can advantageously minimize the volume(s) or amounts of gas used while enabling relatively accurate flow control. The hydrogen/deuterium gas supply manifold can enable predetermined, controlled quantities of gas to be provided into the reaction chamber of the electrolysis system, according to an exemplary embodiment. The carrier gas supply manifold can enable predetermined quantities of carrier gas (e.g., argon) to be provided during start-up and maintenance periods into the exemplary reaction chamber, according to an exemplary embodiment. The reaction gas manifold can enable reactant gases to be temporarily stored and periodically extracted as a saleable and profitable resource, according to an exemplary embodiment. According to an exemplary embodiment, extracted reactant gas can be permanently stored and/or maintained according to applicable standards and regulations using formal methods that are not part of the present disclosure. According to yet another aspect of the present disclosure, the gas measurement and evacuation manifold can enable quantities of gases from the reaction chamber and other sections of the gas handling system to be extracted and analyzed as needed during start-up and maintenance periods, according to an exemplary embodiment. Gases extracted through this manifold during maintenance can also be stored and maintained according to applicable standards and regulations, according to an exemplary embodiment.

These and other aspects, features and advantages according to an exemplary embodiment of the present disclosure are provided by an exemplary modular gas electrolysis apparatus that can include the cooled header for electric connections or couplings, exemplary anode, exemplary cylindrical, co-disposed cathode, exemplary coaxial heat exchanger around the reaction chamber and rugged, gas manifolds, according to one exemplary embodiment.

Moreover, to support industrial applications the exemplary electrolysis apparatus preferably can include an improved, modular electronic control circuit, according to an exemplary embodiment. The improved circuit can provide further advantages in addition to functionality of the electronic control circuitry described earlier in U.S. Pat. No. 6,248,221 B1, the contents of which are incorporated herein by reference in their entirety.

According to one exemplary embodiment, the gaseous electrolysis apparatus can include: a header including at least one electrical connector or coupling; an anode with internal heater; a cathode disposed within a reaction chamber; a heat exchanger in proximity to the reaction chamber, the reaction chamber coupled to the header, the heat exchanger configured to remove heat from a surface of the reaction chamber; a gas handling system mechanically coupled to the reaction chamber; and an electronic control circuit or subsystem (ECC) electrically or electronically coupled to the gas handling system and to the header.

The detail below supports further understandings to bridge the gaps between laboratory research and a commercially useful device for producing heat energy at levels of practical interest by providing additional clarity into alternatively useable embodiments and characteristics of the alternative embodiments.

According to one exemplary embodiment, the header can include a high pressure gasket and at least one electrical connection or coupling configured to enable opening and closing of the gaseous electrolysis apparatus or the replacement of parts.

According to one exemplary embodiment, the gaseous electrolysis apparatus can include a header, including at least one feedthrough to a header cooling chamber proximate to the header.

According to one exemplary embodiment, the header can include feedthroughs, where at least one feedthrough is optimally built so that a pressure side of the feedthrough can be oriented towards the inside of the reaction chamber; where at least one feedthrough is welded into a thermal plate, where one end of the feedthroughs extends beyond the header for connection or coupling with the electronic control circuit or subsystem; or wherein at least one feedthrough can include a threaded coupling.

According to one exemplary embodiment, the gaseous electrolysis apparatus can include a header that includes at least one of: a cooled header which is a coolant cooled header; a liquid cooled header; a water cooled header; or a gas cooled header; or wherein the cooling apparatus can include at least one of: a header cooling chamber; a cooling manifold; a jacket; a cooling jacket; a fin; a cooling fin; a manifold; a fixture; a cooling fixture; a plate; a thermal plate; a header cooling jacket; a header cooling fixture; a cooling manifold; or a water jacket adapted to maintain a sufficiently low temperature for a gasket and at least one feedthrough.

According to one exemplary embodiment, the header can include at least one of: a temperature measurement device to monitor temperature of the header or a cooling apparatus; a gas detector configured to detect any leak through the header and that includes a gas volume coupled to the header; or a microwave loop antenna for irradiating a volume within the reaction chamber that includes a ceramic case, where the ceramic case comprises a ceramic tube, a stabilizer to provide mechanical stability, or a protector.

According to one exemplary embodiment, the header can include at least one insulator configured to electrically isolate, minimize the volume where gas could reside, and provide mechanical support for the anode, cables, electrical connections or couplings, or internal reaction chamber components, and wherein at least one insulator can include at least one of: a ceramic insulator; a plurality of ceramic insulators; a robust, thick ceramic insulator; or a reflector or a reflective heat shield or a baffle on a lower surface in order to reflect heat into the reactor volume.

According to one exemplary embodiment, the gaseous electrolysis apparatus can include an anode coupled to at least one electrical connector or coupling in the header.

According to one exemplary embodiment, the anode can include a heater coupled to an electrical connector or coupling in the header and thus coupled to the electronic control circuit.

According to one exemplary embodiment, the gaseous electrolysis apparatus can include co-disposed within the reaction chamber a cylindrical, coaxial cathode that is of modular design or is configured to be a removable unit and that is electrically coupled, connected or grounded to the reaction chamber.

According to one exemplary embodiment, the cathode can include at least one of: an outer metal sleeve; an outer metal stainless steel sleeve; or be bounded by a base, a top, or with a ceramic insulator endcap(s) that contain a plurality of holes for reaction gas to escape.

According to one exemplary embodiment, the cylindrical, co-disposed cathode within the reaction chamber can include a consolidated metal powder of modular design or a

material that includes metal powder or metal particles, wherein the cathode material can be at least one of: a shape to provide for high packing density and enable consistent cathode industrial production; include a spherical powder/particle of a small medial size on the order of a few microns, with a tight Gaussian distribution around the medial size; a consolidated metal powder with high specified density or have a low value of porosity determined to be a value between 0 and 20%; or comprise, consist, or consist essentially of a single or multiple elements selected from the group consisting of Fe, Ni, Cu, Mo, Cr, Co, Mg, Ag, and W.

According to another exemplary embodiment, the edges of this reaction material part facing the anode can be tapered or curved to reduce the concentration of electric fields between the anode and cathode.

According to one exemplary embodiment, the heat exchanger can include a co-axial heat exchanger or be co-disposed around the reaction chamber and enable a thermoelectric generator to be disposed also around or about a portion of the surface of the reaction chamber, or a system of multiple thermocouples to be disposed around or about a portion of the surface of the reaction chamber.

According to one exemplary embodiment, the heat exchanger can include at least one of: a flash boiler enabling a weight of coolant to be minimized for a mobile or transportable applications; a flash boiler; a relatively low volume boiler; a plurality of spray nozzles to cool at least one portion of the reaction chamber; a boiler configured to provide coolant to at least one portion of the reaction chamber; a thruster port configured to provide pressure output; or a thruster configured to produce an inert gas/steam output.

According to one exemplary embodiment, the plurality of spray nozzles can serve to at least one of: cool the surface of the reaction chamber uniformly; cool the surface of the reaction chamber to at least several tens of degrees below the average temperature within a cathode of the gaseous electrolysis apparatus; or provide a controlled release of a predetermined amount of coolant, wherein the coolant can include a mist of coolant, liquid, gas, or water.

According to one exemplary embodiment, the heat exchanger can include at least one of: a steam pressure port and steam pressure regulator configured to maintain a sufficiently high steam pressure for at least one application, e.g., to make electricity; a configuration to receive coolant for increased system efficiency during long periods of operation from the header and/or gas handling system cooling chambers; or a coolant supply including high pressure water feed pump to provide coolant to the header's cooling manifold, cooling chambers of the gas manifold system and spray nozzles of the heat exchanger.

According to one exemplary embodiment, the gas manifolds can include, as an enhanced safety measure, a plurality of valves that are normally by default closed, but are configured to be opened and closed to move gas into and out of the reaction chamber.

According to one exemplary embodiment, the gas manifolds can include at least one of: a cooling chamber or water jacket to provide cooling for gas tubing and pipes connected or coupled to the reaction chamber; minimal volumes or amounts of gas external to the reaction chamber where heat is produced, enabling relatively accurate flow control and addressing safety concerns; tanks or containers whose known volume enables small quantities of gas to be determined by calculating pressure, temperature and volume before gas is transferred into or out of the reaction chamber; an acoustic sensor/electronic interface, gas detector, or reaction product sensor, configured to manage gas output with a separator valve and facilitate estimation of quantities of reactant gas being processed; a container configured to temporarily store the reactant gas, and periodically permit extraction from the container; a low-voltage, gas compatible valve, pressure sensor and temperature sensor connected or coupled to the electronic control circuit/subsystem; purge ports which can be used to evacuate gases manually from their containments; or a safety tank/bottle for overpressure protection for transfer through a safety valve or rupture disc of gas from the reaction chamber.

The electronic control circuit (ECC) can include components that enable real-time set-up, control, recording and analysis of apparatus operations, further bridging between laboratory experiments and a commercially useful device. According to one exemplary embodiment, the ECC can include at least one of: an electrolysis apparatus controller subsystem configured to measure at least one sensory output of a plurality of field sensors, record at least one sensory output as sampled data, or store the sampled data on an external device; an electrolysis apparatus controller subsystem configured to store in an off-board, permanent memory using a processor board port or universal serial bus (USB) port coupled to the control system, sampled data that can be analyzed and further configured to train and optimize control system parameters contained in software of the electrolysis apparatus controller subsystem; or an electrolysis apparatus controller subsystem configured to create analog or digital signals to be sent to or from a plurality of field control devices, according to at least one control algorithm output in software, in order to control the state of the reaction chamber as indicated by input devices comprising thermocouples, pressure sensors, and/or an acoustic sensor.

According to one exemplary embodiment, the ECC can include at least one of: series and parallel precision resistors or a precision non-contact probe configured to measure electric current flowing across the reaction chamber; a bipolar (±), variable, high voltage power supply configured to vary the voltage level from the power supply or to adjust the biasing on a transistor type component inside the power supply or to adjust the power supply output resistor network to create a desired level of voltage and current across the anode and cathode; a power supply for the electric heater located inside the anode of the reaction chamber; an electronic separator valve subsystem or acoustic interface that facilitates estimation of quantities of reaction gas processed; a thermocouple measuring temperature inside the reaction chamber; a pressure sensor measuring pressure inside the reaction chamber; a microwave generator coupled to a robust loop or other microwave antenna to provide a high-frequency electromagnetic field inside the reaction chamber; an electric interface with heat exchanger cooling valves to support the reaction chamber cooling process; or a nuclear radiation detector configured to sense a radiation level, analyze or compare the sensed radiation level to predetermined thresholds, store radiation level information and generate a notification.

According to one exemplary embodiment, the ECC can include at least one of: a single-board computer that can include a processor chip with accessory and supporting circuitry, a controller area network (CAN) transceiver chip or port, at least one USB port, an RS 232 serial port, a general-purpose input and output (GPIO), an ethernet physical (PHY) input or an output, interface or port connector and a video graphics association (VGA) display or other output port; an external memory; an analog input extender board; an analog output extender board; a control area network

including a digital output module, a digital input module, an analog output module, or an analog input module; a Linux, a Unix, an embedded Linux or Unix, or other operating system (OS); equipment power supplies including isolated 12 and 24 volt DC power supplies; a video display screen or other output device; or input keyboard.

According to one exemplary embodiment, the ECC can include software instructions, which, when executed on a processor can perform functions to at least one of: provide an automated system pre-startup sequence; provide automated charging of mixing chamber and/or reaction chamber; control reaction chamber dynamics during startup and/or operation; provide automatic fail-safe shut down; provide long term, automatic adjustment of control system output signals to the process to facilitate the empirical discovery of the signal levels required to initiate, control or maintain a desired reaction; provide for output or display of process measurements for real-time human observation; enable a human interface with the process through display of interactive video screens where an operator initiates or terminates a process or adjusts specified control system parameters to change the speed, nature or intensity of a process; provide detection of significant system events or combinations of events, measured from at least one sensory input field device, or provide a corresponding threshold table; or provide a collection of data from at least one, or all sensory input devices, sampled or stored on an external device connected or coupled through a processor board, port or USB port.

According to one exemplary embodiment, the ECC can include software instructions, which, when executed on a processor, can perform functions to at least one of: accept one or more input(s) and calculate one or more operating output(s) to control hardware automatically and in such a manner to satisfy engineer-specified setpoints; adapt to varying system conditions so as to appropriately adjust internal operation parameters and where the same specified operations or functions comprise a multiple-input and multiple-output (MIMO) control process; adapt to varying system conditions using a proportional-integral-derivative (PID) controller, a modified PID controller, a least means squares (LMS) algorithm, a Gradient Search by a Steepest Descent algorithm, a Box-Jenkins algorithm, a Hammerstein-Wiener System estimator, a Radial Basis Network system, or a Principal Component Analysis (PCA) system.

According to one exemplary embodiment, the ECC can also include software instructions, which, when executed on a processor can perform functions to at least one of: perform system identification (ID) processing using at least one system ID process; efficiently generate informative data for fitting a mathematical model of the system; provide sufficiently rapid response time for precision control of discrete quantities of deuterium, hydrogen and carrier gas molecules supplied to the reaction chamber during apparatus operation; and control a reaction chamber exhaust valve, configured as a fail-safe to exhaust pressurized reaction chamber gases in event an input device measurement exceeds a safety threshold.

BRIEF DESCRIPTION OF THE DRAWINGS

The embodiments are described with reference to the drawings in which like elements are denoted by like or similar numbers and in which:

FIG. 1, which includes FIGS. 1A-1D, collectively illustrates an exemplary embodiment of a modular header according to the present disclosure which can provide a relatively cool temperature environment for exemplary electrical feedthroughs to support long-term operation of the system, according to an exemplary embodiment.

FIG. 1A indicates a configuration for an exemplary reaction chamber and flange for attachment of the modular header of the exemplary embodiment.

FIG. 2, which includes FIGS. 2A-C, illustrates an alternative exemplary embodiment according to the present disclosure of the manner in which a heat exchanger module facilitates and controls a thermal gradient through the cathode module and the wall of the reaction chamber module. The thermal gradient, according to an exemplary embodiment, facilitates production of output heat as a continuous process, extracting the reaction heat as, for example, steam for application to an external electric generator, space heater or a mechanical process.

FIG. 2A depicts an exemplary cutaway one half view of an exemplary reaction chamber illustrating an exemplary thruster port for inert gas or steam, and illustrates the heat exchanger or boiler, according to an exemplary embodiment.

FIG. 2B depicts a zoomed in portion of the exemplary reaction chamber of FIG. 2A, according to an exemplary embodiment.

FIGS. 3A-E depict various exemplary illustrations of exemplary gas manifold subsystems according to an exemplary embodiment.

FIG. 3A is an illustration, according to an exemplary embodiment, which is helpful in understanding connection or coupling of the exemplary gas manifolds to the reaction chamber so as to minimize gas volume external to the reaction chamber and remove heat from the gas manifolds according to an exemplary embodiment.

FIG. 3B depicts an exemplary top perspective view of an exemplary hydrogen/deuterium gas supply manifold, according to an exemplary embodiment.

FIG. 3C depicts an exemplary perspective view of an exemplary carrier gas supply manifold, according to an exemplary embodiment.

FIG. 3D depicts an exemplary perspective view of an exemplary reaction gas product collection manifold, according to an exemplary embodiment.

FIG. 3E depicts an exemplary perspective view of an exemplary gas measurement and evacuation manifold according to an exemplary embodiment.

FIGS. 4A-B collectively form an exemplary embodiment of detailed schematic diagrams of the exemplary improved electronic control circuit according to an embodiment of the present disclosure.

FIG. 4A provides an electrical control connectivity diagram, according to an exemplary embodiment.

FIG. 4B provides an example process and software architecture, according to an exemplary embodiment.

FIG. 5 depicts an example mechanical steam controller system connected or coupled to the exemplary improved electronic control circuit according to an embodiment of the present disclosure.

FIGS. 6A-6F, collectively referred to as FIG. 6, demonstrate an example construction of components for an example reaction chamber, header, high temperature feedthroughs, anode, cathode part, and separator valve/acoustic sensor, respectively, according to example embodiments of the present disclosure.

DETAILED DESCRIPTION OF THE PREFERRED EMBODIMENTS

An exemplary embodiment of the disclosure sets forth improvements to an exemplary gas electrolysis system

enabling such a system to bridge the gaps between laboratory apparatuses and a commercially useful device, and to be manufactured by industry. The present disclosure sets forth features enabling these exemplary industrialized systems to have large power handling capabilities and address well-known problems with the state of the art and conventional solutions, such as the possibility of hydrogen recombination with oxygen, the relatively slow loading of gas into cathodes, inefficient designs and potential dangers of loaded, pressurized bulk material, while being regulated for ease of operation.

Information in the previous and the following paragraphs includes preferred embodiments and various alternative exemplary embodiments for the disclosure, operational procedures for the disclosure and information on modes of practicing the disclosure. The exemplary improvements discussed, moreover, support key design parameters of the exemplary apparatus, such as, e.g., but not limited to, sufficient cathode reaction material for planned power output, relatively high operating temperature and pressure, and minimal volumes in the gas manifolds for safety. This information enables the subject matter of the claims (35 USC § 112) and teaches/shows how to use this invention (35 USC § 112(a) and pre AIA 35 USC § 112). FIGS. 1-6, in accordance with 37 CFR § 1.121(d), include a depiction of each claimed component.

INTRODUCTION

In complementing previous advancements, the basis for the present disclosure is to bridge the gaps between laboratory experiments and a commercially useful device for producing energy at levels of practical interest and that can be manufactured by industry. An objective is to demonstrate that hundreds of watts of heat can be produced in cubic centimeters of specific reaction materials. For practical applications, this can require apparatuses that sustain on the order of 10 to the exponent 16 nuclear reactions per second, assuming each reaction nets several million electron volts (MeV) of energy. The exemplary design disclosed is based upon use of pressurized hydrogen and/or deuterium gas and includes, for example: an example reaction chamber or reactor whose purpose is to enclose active high temperature and pressure system components; an example closure, otherwise known a "header," whose purpose is to seal the reaction chamber from the outside atmosphere and provide electrical connections between components within the reaction chamber and external components; an example heat exchanger to extract heat from the outer surface of the reaction chamber; an example gas handling system that can include separate gas manifolds to control gas flow in and out of the reaction chamber; and an example electronic control circuit (ECC) electrically connected to, or coupled to, the header and gas handling system. One of the example gas manifolds enables either deuterium or hydrogen, or both, to be provided in measured amounts to the reaction chamber. One of the example gas manifolds enables reactant gas to be stored and periodically extracted from the system. The example header can contain a cooling apparatus to enable electrical feedthroughs to be cooled to increase their operational lifetime and can provide mechanical support for other components within the reaction chamber. Internal to the reaction chamber, an embodiment of the design can include an anode, which can include an internal heater, and an example removable cylindrical cathode around the anode. The electronic control circuit (ECC) can enable real-time set-up, control, recording and analysis of apparatus opera-

13

14

tions. The apparatus is a complex system comprising (or including) these key components.

Specific and substantial utility (as required by 35 USC § 101) is ample, and includes using the disclosed system or apparatus to first provide for laboratory applications and investigation and testing, in an example embodiment. The example system of the claimed invention has various example potential uses, including use in the study of, and testing of, electrolysis reactions, and cooling and heat transfer via exemplary heat exchangers and the like of various materials of exemplary embodiments. The capability of the example header to cool example feedthroughs can advantageously enable longer periods of operation between maintenance, in an exemplary embodiment. The capability of the header to integrate internal reactor components into the lower side of the header supports apparatus maintenance, in an exemplary embodiment. The example co-disposed heat exchanger around the example reaction chamber has an exemplary purpose of enabling useful power to be extracted in an efficient and safe manner, and is, e.g., removable for maintenance. The example heated anode has an exemplary purpose of enabling temperature of reaction material within the cathode to be raised above its Debye temperature. The heated anode within the reaction chamber and heat exchanger can support thermal diffusion of deuterium and/or hydrogen gas into the cathode reaction material, in an exemplary embodiment. The thermal diffusion gradient, along with electric fields and gas pressure, are considered to be highly important in promoting gas transport into and through the reaction material, in an exemplary embodiment. The example cathode in one embodiment can include a modular, removable component and can be encased in a metal supporting sleeve to facilitate removal/replacement during maintenance. An example purpose of one of the gas manifolds is to enable control of the necessarily very small amounts of hydrogen and/or deuterium gas that can be provided to the reactor. One of the other manifolds is designed to enable reaction product gases to be quantified and stored, in an exemplary embodiment. The example ECC enables control of key physical processes and parameters within the reactor, such as, e.g., but not limited to, anode temperature, gas pressure, electric field strength, and type and rate of gas diffusion, as well as controlling supporting components, etc. A person of ordinary skill in the art would, therefore, appreciate reasons the invention is useful based on these and other benefits. These capabilities support the need for long operation periods, ease of maintenance and safe operations. An example modular systems approach for component design can enable many numbers of industries to manufacture, install, repair and otherwise support these objectives.

Early concepts related to design of the apparatus are provided in a paper on "Critical Factors in Transitioning from Fuel Cell to Cold Fusion Technology," by T. McGraw and R. Davis in August 1998, which depicts the simplistic arrangement between the anode and cathode and indicates the importance of a long cathode with relatively large surface area, a thermal gradient combined with gas pressure and electric field in loading the cathode, a collection bottle for reaction product helium, and of safety as a primary concern. Additional information related to the design is presented in U.S. Pat. No. 6,248,221 B1, "Electrolysis Apparatus and Electrodes and Electrode Material Therefor," by Davis et al., Jun. 19, 2001, which discloses a consolidated cathode reaction material including or comprising nanocrystalline particles, a porous insulator reaction vessel between the anode and cathode, a microwave type of starter/

initiator and an electronic control circuit that controls electricity between the anode and cathode. Concepts related to design of the apparatus are presented in the draft of a paper on "Key Issues Related to Industrialization of LENR-Based Space Propulsion," by M. Chawla and R. Davis that was developed in April 2016 for the Space Technology & Applications Industrial Forum (STAIF) and emphasizes that fuel quantity supplied to the reaction material needs to be controlled to limit reaction rate in the reaction material and that heat must be removed efficiently. An early gas loading concept is discussed in patent application WO 95/20816, "Energy Generation and Generator by Means of Anharmonic Stimulated Fusion," by S. Focardi et al., Aug. 3, 1995, that reports on use of electricity through a high-temperature coil to load and heat reaction material above its Debye temperature to produce fusion between hydrogen and deuterium and on methods to initiate the fusion reactions. By comparison, and demonstrative of the novelty of the disclosure under 35 USC § 102 and the nonobviousness of the disclosure under 35 USC § 103, these sources of information do not disclose a header according to the disclosure, where feedthroughs are cooled and that provides for integration of internal reactor components; a heat exchanger that is modular and removable; a modular anode with an internal heater; a modular, removable cathode encased in a metal supporting sleeve; gas manifolds that are able to control very small amounts of gas to the reactor and quantify amounts of reaction gases produced; an electronic control circuit that enables control of the very small amounts of hydrogen and/or deuterium gas provided to the reactor and the quantification of reaction product gases; nor the means to remove heat efficiently, according to an exemplary embodiment. These elements of the instant application described herein provide significant differences and depend upon unique technical concepts in the design and were not previously self evident to a person having ordinary skill in the art. The disclosed invention is novel. This disclosure was also made by parties in the joint research activity that includes the same parties for the Critical Factors and Key Issues papers and U.S. Pat. No. 6,248,221 B1. There has been no known teaching, suggestion or motivation by others to combine the above or other references in the manner described in this application.

Internal to the example reactor, the apparatus is viewed to operate by "gas electrolysis" or "gaseous electrolysis", as this description most effectively captures the essence of electrochemical processes in the volume between the anode and cathode and that involves a mixture of molecules, ions and electrons, pressurized hydrogen and/or deuterium gas, elevated temperatures and strong electric fields. Supporting data are provided in a great number of scientific works since the late 1800s to the present day regarding electrical conduction through hydrogen gas. See the paper, "On the Electrolysis of Gases," by J. J. Thomson and pages 270-4 in Chapter VIII and pages 293-4 in Chapter IX in the text on "Theory of Gaseous Conduction and Electronics," by F. A. Maxfield and R. R. Benedict. High voltage breakdown or avalanche discharge through the gas is not desired. The Paschen curve for hydrogen and Townsend criterion can be used to ensure that sufficiently low voltages and high gas pressures are used to prevent breakdown. See pages 188-190 in "Introduction to Electrical Discharges in Gases," by S. C. Brown.

Physical similarities exist between concepts for gas-based and liquid-based LENR apparatuses whose understanding can thereby support the transition from liquid to gas LENR systems. Each contains cathodes where reactions can be

15

16

made to occur, anodes, electrolytes (i.e, gas or liquid) and direct (dc) drive currents. A liquid-based system is concerned with anions and (e.g., D^+ and/or H^+) cations, their movement in a liquid electrolyte and cathodic interactions. Gas-based concepts, by comparison, are concerned with mechanisms that can form positive ions from (e.g., deuterium and hydrogen gas) molecules, their movement to the cathode and cathodic interactions. Several ion forming mechanisms (elastic, excitation, ionization) can be considered, but the most important is due to collisions of thermal electrons with gas molecules. Ionization cross sections vary in a non-linear manner. Energies of scattered electrons are frequently increased in the scattering process. The resulting mixture can contain many different species of ions and molecules that interact with various cross sections as described in "Cross Sections and Swarm Coefficients for H^+, H_2^+, H_3^+, H, H_2 and H^- in H_2 for Energies from 0.1 eV to 10 keV," by A. V. Phelps. The positive species can be accelerated toward the cathode at different rates determined by their mass and charge.

FIGS. 1A-1D, collectively referred to as FIG. 1, illustrate an exemplary embodiment of a modular header according to the present disclosure which can provide a relatively cool temperature environment for exemplary electrical feedthroughs to support long-term operation of this apparatus.

FIG. 1A illustrates an embodiment of a portion of an exemplary reaction chamber 122 (described further with reference to FIG. 1D) including a flange portion with a plurality of openings for coupling to an adjacent header 120 of the exemplary apparatus; also shown are exemplary ports for coupling the reaction chamber to one or more gas manifolds as described further herein.

FIG. 1B depicts an exemplary cut-away top view of an exemplary header assembly 100 including various exemplary feedthroughs 102-112, including an exemplary copper wire anode connection 102, anode heater wire connections 104 and 106, an exemplary 4-wire connection 108, an exemplary microwave antenna connection 110, and a thermal sensor connection 112 as may be provided in a preferred embodiment of the header. The anode in a preferred embodiment can include a heater, such as, e.g., but not limited to, a cartridge-type heater encased in the anode (for reference, see, e.g., Dalton Electric Heating Company model W6C120).

Also shown in FIG. 1B is an exemplary header 120, exemplary reactor or reaction chamber 122, exemplary header cover or cooling plate 124, exemplary thruster/steam port 218, and partial view of cathode 208 and anode 202 with internal heater discussed further below, in an exemplary embodiment.

An exemplary embodiment of the exemplary header 120 for the exemplary electrolysis apparatus 100 can enable the exemplary reaction chamber 122 to be opened and closed for replacement of internal parts. The exemplary cathode 208 can be replaced when the cathode no longer operates efficiently as may be evident when anode 202 current or anode-to-cathode voltage does not meet allowed threshold values, or the system's coefficient of performance (COP) is determined to be insufficient. The exemplary header 120 can include an exemplary header cover, closure gasket and/or flange that can serve as an exemplary main seal for the apparatus and can be designed to maintain specified mechanical integrity and reliability during many temperature and pressure excursions, according to an exemplary embodiment. Standard operating pressure and temperature within the exemplary reaction chamber 122 are hundreds of

pounds per square inch and hundreds of degrees centigrade, according to an exemplary embodiment. During exemplary maintenance and startup periods, according to an exemplary embodiment, the bottom of the exemplary header 120 along with the inside of the exemplary reaction chamber 122 can be required to be exposed to high-temperature, pressure, and vacuum cycles to remove oxygen and other gaseous impurities from the reaction chamber 122. An exemplary carrier gas, e.g., argon, can be introduced to physically flush the system, according to an exemplary embodiment. During startup, pressure in the exemplary reaction chamber 122 can then be required to increase in steps as, first, deuterium and, then, hydrogen are made to enter the reaction chamber 122, according to an exemplary embodiment. According to an exemplary embodiment, the reaction chamber 122 can be required to operate at high pressure and temperature over long periods of time, wherein additional gas can be periodically introduced to maintain the heat energy producing process.

According to one important aspect of this disclosure, the header 120 can contain a thermally cooled plate or header cover 124 through which one or more feedthroughs 102-112 can be mounted, according to an exemplary embodiment. The exemplary feedthroughs 102-112 can be standard, commercially-available, high voltage, power and coaxial electric feedthroughs, and can include electrical conducting members and insulating material such as alumina or glass, according to an exemplary embodiment. These feedthroughs 102-112 can be manufactured by metallization, high temperature consolidation and/or vacuum brazing methods known to those skilled in the art of hermetic ceramic-to-metal sealing technology, according to an exemplary embodiment. See, for example, U.S. Pat. No. 4,174,145, "High Pressure Electrical Insulated Feed Thru Connector," Nov. 13, 1979; U.S. Pat. No. 4,593,758, "Hermetically Sealed Insulating Assembly," Jun. 10, 1986; and U.S. Pat. No. 5,273,203, "Ceramic-to-Conducting-Lead Hermetic Seal," Dec. 28, 1993, the contents of which are incorporated herein by reference in their entirety. For reference, see, e.g., Solid Sealing Technology website at www.solidsealing.com. The exemplary feedthroughs 102-112 can, in an exemplary embodiment of the disclosure, maintain their integrity during the above described variable and long, elevated pressure and temperature environment, according to an exemplary embodiment. The use of the exemplary feedthroughs 102-112 in headers/isolators for these types of apparatuses containing high-pressure hydrogen at high temperature has never before been previously proposed, according to an exemplary embodiment. In the relatively cool temperature environment provided by the cooling plate 124, however, the feedthroughs 102-112 can be able to operate consistently to support long-term operation of the system, according to an exemplary embodiment.

FIG. 1C depicts an exemplary cutaway view 130 of exemplary feedthroughs 102-112 in an exemplary configuration. The feedthroughs 102-112 can be permanently mounted through the exemplary thermal cooling plate 124, for example, by tungsten inert gas (TIG) welding, in an exemplary embodiment. In a preferred embodiment, the feedthroughs can be built so that the pressure side 128 of each feedthrough 102-112 can be towards the inside of the reaction chamber 122 (not shown), in an exemplary embodiment. Connections or couplings, in an exemplary embodiment, can be made from the pressure side of the feedthroughs to the anode, heater, microwave antenna, thermal sensor or other desired electrical device(s) as may be located within the exemplary reaction chamber 122. Each feedthrough 102-112 can also be connected or coupled with external electronic control circuitry (not shown) above the header cover/cooling plate 124, in an exemplary embodiment. Alternatively, in one exemplary embodiment, the feedthroughs 102-112 can contain an exemplary circular disk flange header cover or cooling plate 124, which may be continuous around the circumference of the feedthroughs 102-112, to enable the feedthroughs 102-112 to be mounted as shown in FIG. 1D. In an alternative configuration, the feedthroughs 102-112 can be mounted with exemplary pressure gaskets or seals through the thermal cooling plate 124 so that the feedthroughs 102-112 can be replaced when damaged, in an exemplary embodiment. Alternatively, in another exemplary embodiment, each feedthrough 102-112 can first be welded into a screw-type metal socket or ferrule which can be installed in the mounting plate 124 with a suitable gasket. Again, the exemplary pressure side 128 of each feedthrough 102-112 can be directed towards the inside of the reaction chamber and connections or couplings can be made from the pressure side of the feedthroughs 102-112 to elements within the reaction chamber 122, in an exemplary embodiment. In relatively low pressure configurations, it is considered possible, in another exemplary embodiment (not shown), for the feedthroughs 102-112 to be mounted with pressure gaskets or seals through the thermal cooling plate with their pressure side reversed so that the pressure gaskets or seals can be more easily able to be replaced. In this latter exemplary case, the pressure side of each feedthrough can be towards the outside of the reaction chamber 122 and connections or couplings can be made from the lower side of the feedthroughs 102-112 to elements within the reaction chamber 122 (not shown). All such alternative configurations can be thereby considered to be within the scope of this patent.

The lower side of the header 120 can contain exemplary robust, thick insulators that can electrically isolate and provide mechanical support for the electrical connections or couplings and can, e.g., for improved operation and safety, ensure that gas in the reaction chamber is concentrated near the cathode and minimize the volume where gas could reside, in an exemplary embodiment. The exemplary upper insulator can provide electrical isolation between connections, in an exemplary embodiment. It is particularly important, for example, in an exemplary embodiment, to electrically isolate the anode 202 connection or coupling 102 from all structural, control and sensory elements where stray currents can be induced in system grounds, earth grounds, power supply voltage sources or other control system voltage sources. The anode-cathode current, therefore, can be measured and thereby controlled with maximum accuracy, in an exemplary embodiment. The exemplary lower insulator 1402 (see FIG. 2A) can provide mechanical support for a microwave antenna, anode 202 and thermal sensor, in an exemplary embodiment. The lower insulator 1402 can also contain an exemplary reflective heat shield or baffle on its lower surface to reflect infrared energy (heat) back down into the reactor volume, in an exemplary embodiment. In an embodiment, these insulators can be attached to the header 120 closure flange. The insulators, anode, microwave antenna, thermal sensor, etc. can thereby be able to be installed and removed as an integral unit, in an exemplary embodiment. In alternative embodiments, the top of the header 120 can be opened and closed separately from the insulators, subsequent to installation of the insulators with anode 202, microwave antenna, thermal sensor, etc. into the reaction chamber 122.

According to an exemplary embodiment of this disclosure, the header 120 can contain a cooling manifold or water

jacket that can provide cooling for the thermal cooling plate. See FIG. 1D. The cooling plate 124 with attached feedthroughs 102-112 and installed conductors can thereby be required to make thermal contact with the cooling manifold. The cooling plate 124 can be attached and/or mechanically coupled by a strong, continuous and/or leak-proof weld to prevent gas from escaping the reaction chamber 122 through the header 120, in an exemplary embodiment. Alternatively, for lower pressure applications the cooling plate 124 can be attached, e.g., with suitable bolts and/or a gasket that can prevent gas from escaping from the reaction chamber 122, in an exemplary embodiment. In such a case, maintenance may be performed by removing the cooling plate 124, replacing a damaged feedthrough 102-112 or other component and then re-assembling the header 120, in one exemplary embodiment. Exemplary ceramic tube insulators, in an exemplary embodiment, can be installed around the conductors extending through holes in the cooling manifold to provide electrical isolation and minimize the volume where gas could reside. The flow design of the exemplary cooling manifold advantageously can provide a large surface area and liquid volume for transfer and removal of heat from the header 120, resulting in lower local temperature both adjacent to the feedthroughs 102-112 or feedthrough assemblies and for the header seal, in an exemplary embodiment. Furthermore, an electronic temperature measurement device (e.g., thermocouple) can be mounted on the cooling plate 124 where the temperature can be monitored and maintained at a relatively constant value during long, high temperature operational periods, in an exemplary embodiment. The amount of heat removal can be adjusted using temperatures from the thermocouple and coolant flow rate to change the coolant's volume through the cooling manifold.

Design of the header 120 can include several exemplary safety features, in an exemplary embodiment, in addition to mechanical integrity of an exemplary embodiment of the present disclosure. The closure mechanism, in an exemplary embodiment, bolting the header to the reaction chamber flange, can be made to operate so that required torques on the bolts can be checked. Gas pressure that could decrease due to a leak through a feedthrough 102-112 or accumulate at the edge of the header flange can also be monitored by a separate electric gauge, in an exemplary embodiment. This can be made possible, both by a sealed cover over the feedthroughs 102-112 attached to the top of and is part of the header and by a removable, wide vertical gasket that can surround and lap across the header flange and reaction chamber flange and that can be tightly secured by an exemplary separate mechanical support band around the header flange and the reaction chamber flange, in an exemplary embodiment. Electrical conductors extending from the feedthroughs 102-112 to the external electronic control circuitry can be made to pass through the gas safety cover, in an exemplary embodiment. The cover can press fit to the main part of the header 120 and can be easily removed for maintenance, in an exemplary embodiment. Either an electronic pressure or gas sensor/detector can be located within the safety cover and can be used to detect any gas leak through the header 120 and thus can be configured to electronically provide an alert signal to the apparatus operator.

Referring specifically to FIG. 2A, according to an exemplary embodiment, the coaxial cathode 208 where heat can be produced can be located within the reaction chamber 122 and can touch the reaction chamber's inner wall 204 to facilitate heat conduction through the reaction chamber wall 122. An aspect of an exemplary embodiment can be to enhance the thermal diffusion gradient between the anode **202** through the cathode **208** reaction material and reaction chamber wall **122**, and, can thereby, facilitate thermal diffusion of gas through the cathode **208** over long periods of operation, according to an exemplary embodiment. Heating up the anode in the center of the cathode heats up the inner part of the cathode and the gas moves with the thermal gradient, and through the cathode by thermal diffusion, according to an exemplary embodiment. Also see cutaway view **230** of FIG. 2B.

The cathode in a preferred embodiment can be a hollow-shaped cylinder, with a central cavity configured to receive the anode. The cathode in a preferred embodiment can be encased by an exemplary outer metal (e.g., stainless steel) sleeve **212** and can be bounded at its base and top with exemplary ceramic insulator endcaps **1404** and **1406**. The cathode can contain a porous insulator reaction vessel **206** on its inner surface facing the anode. The cathode is electrically grounded and can be co-disposed (or optimally coaxial) in the reaction chamber, according to an exemplary embodiment. In an exemplary embodiment, the upper ceramic end cap insulator **1404** can contain exemplary holes for reaction gases to escape. The cathode in a preferred embodiment can be modular and/or easily removable as a component. An exemplary cathode can include metal powder or metal particles in an exemplary embodiment. According to another exemplary embodiment, the edges of this reaction material part facing the anode can be tapered or curved to help prevent high voltage breakdown between the anode and cathode. In an exemplary embodiment, the particulate material of which this part is constructed can be optionally of a shape to provide for high packing density and enable consistent cathode industrial production. In an exemplary embodiment, the powder/particle size can be a small medial size of about on the order of microns, with a tight Gaussian distribution around the medial size, in one exemplary embodiment. In an exemplary embodiment, the cathode's material can be consolidated metal powder with high theoretical density in an exemplary embodiment. In an exemplary embodiment, the cathode's consolidated material can also have a specified low value of porosity (i.e., the inverse of percent consolidation) and determined to be a value between 0 and 20% in an exemplary embodiment. According to one aspect of the present invention, the cathode's particles can be formed from a single element or multiple elements selected from the group consisting of Fe, Ni, Cu, Mo, Cr, Co, Mg, Ag, and W. In an exemplary embodiment, the cathode can be made of high purity nickel, due to its relatively high abundance on the earth which can support a need for low-cost materials. Additional details on a consolidated cathode, which in an exemplary embodiment can be configured to have a co-axial cylindrical shape, are described in U.S. Pat. No. 6,248,221 B1, issued Jun. 19, 2001. Transport of gas within the cathode can be treated by assuming that deuterium and/or hydrogen can be subject to a superposition of electro- and thermo-transport forces that can cause the gas(es) to move constantly through the reaction material, in one exemplary embodiment.

Another aspect of an exemplary embodiment of the present disclosure can be to provide an exemplary coaxial heat exchanger surrounding the reaction chamber that can provide an exemplary cooling medium (e.g., but not limited to, water) to remove heat from the outer surface of the reaction chamber **122** wall, in an exemplary embodiment. A portion of the annular space surrounding the reaction chamber **122**, in an exemplary embodiment, can form the heat exchanger (e.g., steam generator) **214** of the apparatus as

depicted in cutaway view **200** of FIG. 2A, in one exemplary embodiment. Exemplary thruster/steam ports **218** are further illustrated in FIGS. 2A and 2C, in one exemplary embodiment. The heat exchanger **214** outside the reaction chamber **122** heats up from the reaction chamber's outer surface and the heat in the heat exchanger **214** can heat the water or other liquid, boiling the liquid into steam to turn a turbine, use as thrust (via exemplary thruster port **218**), and/or other heating or power applications as discussed herein, in various exemplary embodiments. The shell and fittings of the heat exchanger **214** (shown in FIG. 2A) advantageously can be made of corrosion-resistant steel, in an exemplary embodiment, for long duration, high temperature operation. According to a preferred aspect of the disclosure, the heat exchanger **214** can be a relatively low volume flash boiler, which can be engineered to provide a mist of water or other liquid coolant to the outer surface of the reaction chamber **122**. Although flash boilers have previously been widely used in other applications, their use in these apparatuses, in an exemplary embodiment, has never before been proposed. Advantageously, the boiler, in a preferred embodiment, can include a sufficient number of spray nozzles **216**, (see, e.g., FIGS. 2B and 2C) to cool the surface of the reaction chamber **122** uniformly several tens of degrees below the average temperature within the cathode **208** of the electrolysis apparatus to support the thermal diffusion process. Cooler water can be provided during start-up, and can support the thermal diffusion process described above, in an exemplary embodiment. An exemplary input for the heat exchanger/boiler advantageously can also be designed to receive coolant that can pass through the header assembly **100** and cooling chambers **302** of the gas manifold system described further with reference to FIGS. 3A-E during long periods of operation for increased system efficiency, in an exemplary embodiment. The exemplary coolant supply can contain one or more high pressure water feed pumps to provide sufficiently high pressure coolant to the cooling manifold of the header, cooling chambers **302** of the gas manifold system, and spray nozzles of the boiler/heat exchanger **214**, in an exemplary embodiment. According to yet another aspect of the present disclosure, the boiler, in an exemplary embodiment, can also connect with steam and water pressure regulators to maintain sufficiently high steam pressure and temperature within the boiler as a result of a high cathode operating temperature set-point, in one exemplary embodiment. Electrically or electronically controllable valves, temperature and pressure sensors can be electrically connected or coupled to an electronic control circuit (ECC) **400** (described further with reference to FIGS. 4A-B) to regulate coolant flow through the header **120** and heat exchanger **214** both to support required temperatures for the reaction chamber **122** and to maintain required thrust/steam output for the system, in an exemplary embodiment. The controller can be designed to maintain the cathode **208** above the Debye temperature of nickel (about 200° C.), according to an exemplary embodiment.

The heat exchanger **214**, in an exemplary embodiment, can be envisioned to supply steam conventionally in a closed-loop configuration to downstream power applications, some of which can be mobile or transportable. The heat exchanger can enable the weight of cooling water to be minimized for these potential applications. The unused steam and condensate can be recovered through a condenser and can be returned by the feed pump to the boiler nozzles and/or header manifold. The system necessarily can include various mechanical steam controllers, pressure valves and/or piping for the different applications, according to an exemplary embodiment. Use in spacecraft power applications can require modifications to this basic design.

According to one exemplary embodiment, it is obvious (FIG. 1D) that an exemplary additional benefit can be provided by adding an exemplary thermoelectric generator comprising an exemplary system of thermocouples around at least a portion of the reactor outer surface so that heat from the reactor can also be used to produce electricity directly, according to an exemplary embodiment. (See for example, the article on "Use, Application and Testing of Hi-Z Thermoelectric Modules" by F. A. Leavitt et al, 2007).

Another aspect of an exemplary embodiment of the present disclosure can include providing an exemplary compact, modular gas handling system for the electrolysis apparatus, in an exemplary embodiment. Specific functions of such an exemplary gas handling system (see FIGS. 3A-E) can include: (1) provide gases to the reaction chamber; (2) enable gases to be extracted from the apparatus during operation and maintenance periods; (3) provide temperature and pressure data used to determine quantities of gases supplied to the reaction chamber, extracted gas product and gas removed during maintenance periods; and, (4) monitor pressures and temperatures to maintain safe operating conditions. From the illustrations of the exemplary embodiment depicted in FIGS. 3A-E, it will be appreciated that the gas handling system can be designed and engineered to be rugged, and occupy an exemplary minimum volume so as to support future mobile and transportable applications. It will also be noted that minimal volume of high pressure gas external to the reaction chamber is an important safety aspect, in an exemplary embodiment. The gas handling system, in an exemplary embodiment, can include commercially-available, low-voltage, gas compatible valves, pressure sensors and temperature sensors connected or coupled to an electronic control circuit/subsystem 400 to, along with operation of the heat exchanger 214, control system operation. The temperature sensors, in an exemplary embodiment, can be calibrated resistive devices that can reflect temperature changes as changes in resistance, which preferably can produce a respective voltage drop across the sensor. Pressure sensors can include a transducer, input power connection or coupling and/or an output measurement monitor that can adjust readings for voltage offset and temperature variations, in an exemplary embodiment. Electric current to valves and sensors can be supplied by the electronic control circuit 400 illustrated in FIGS. 4A-B, in an exemplary embodiment.

The gas handling system (FIGS. 3A-E) of an exemplary embodiment can advantageously include an exemplary four (4) separate gas manifolds that can control gas flow while minimizing gas volume external to the reaction chamber: 1) an exemplary hydrogen/deuterium gas supply manifold 310, 2) an exemplary carrier gas supply manifold 320, 3) an exemplary reaction gas product collection manifold 330, and 4) an exemplary gas measurement and evacuation manifold 340, in an exemplary embodiment. When installed, each manifold can connect or couple as depicted in FIG. 3A to separate input ports of the reaction chamber 122 (see, e.g., FIG. 1A), and each manifold can be constructed as a continuous unit to prevent air from entering the manifolds or the reaction chamber 122, in an exemplary embodiment. According to one aspect of this disclosure, each of the gas manifolds can include, e.g., but not limited to, a cooling chamber or water jacket 302 as shown in FIG. 3A that can provide cooling for gas tubing and pipes connected or coupled to the reaction chamber 122 of FIGS. 1A and 2A for carrying away heat given out by the reaction chamber, in an exemplary embodiment. These water jackets 302, in an

exemplary embodiment, can surround the pipes and tubing so that water through the water jackets can come into contact with the gas tubing and pipes to transfer heat from the tubing and pipes, thereby preventing heat from harming the materials with which the electric valves may be constructed, for example.

FIG. 3B depicts an exemplary perspective view of an exemplary hydrogen/deuterium gas supply manifold 310, according to an exemplary embodiment. FIG. 3B depicts an exemplary embodiment including, e.g., but not limited to, an exemplary cooling chamber 302, exemplary mechanical valves 304, exemplary hydrogen and deuterium gas canisters 311, exemplary gas regulators 312, an exemplary gas mixing and pressurization container with thermal sensor 313, exemplary electric valves 314, and exemplary pressure sensor(s) 315, in one embodiment.

As illustrated in FIG. 3B, the hydrogen/deuterium gas supply manifold 310, in an exemplary embodiment, can enable predetermined quantities of hydrogen and deuterium gas to be admitted into the reaction chamber of the electrolysis system, in an exemplary embodiment. The exemplary gas supply manifold can be designed and engineered to contain a relatively low volume of gas and to provide appropriate quantities to the reaction chamber 122, in an exemplary embodiment. Exemplary separate connections and/or couplings can be provided to high-pressure, high purity gas bottles, such as hydrogen and deuterium bottles, which sources can be controlled by separate mechanically and/or by electrically controlled valves 314. According to another aspect of the present disclosure, the exemplary gas supply manifold 310, in an exemplary embodiment, can be required to operate in short, incremental steps due to the small quantities of gas involved in the heat-producing reactions, in an exemplary embodiment. With hundreds of watts of heat from cubic centimeters of reaction material, only approximately 10 to the exponent 16 molecules per second are needed. For this reason, but advantageously as an enhanced safety measure, the valves can be engineered to be normally closed, in an exemplary embodiment, but then can be opened and/or closed, so as allow gas to move into the reaction chamber 122, in one exemplary embodiment. For reference, see, e.g., Clark Cooper valve model EH40-04A120-HY. Commercially-available valves can typically open and close in 50 to 100 milliseconds (0.050-0.1 seconds). According to yet another aspect of the present disclosure, it should also be noted that the exemplary gas supply manifold can include a small gas mixing and pressurization container or tank 313 whose known volume can enable these small quantities of gas to be determined through pressure, temperature and volume calculations before the gas can be transferred into the reaction chamber 122, in an exemplary embodiment. The manifold can also contain purge ports 316, which can be used to evacuate gases manually from its components as needed.

FIG. 3C depicts an exemplary perspective view of a carrier gas supply manifold 320, according to an exemplary embodiment. Depicted is an exemplary carrier gas supply manifold, including exemplary cooling chamber 302, exemplary mechanical valves 304, exemplary carrier gas canister 321, exemplary gas regulator 322, exemplary electric valve (s) 314, and/or exemplary purge port 316, in one embodiment. The carrier gas supply manifold 320 can control carrier gas emitted into the exemplary reaction chamber 122 of FIG. 2A during start-up and maintenance periods, in an exemplary embodiment. Similar to and as for the hydrogen/deuterium gas supply manifold 310, the carrier gas supply manifold 320 can also be designed to contain a low gas volume and to provide very small quantities of carrier gas (preferably argon) to the reaction chamber, in an exemplary embodiment. The valves of carrier gas operate in short, incremental steps with carrier gas valves normally closed but opened and closed to move carrier gas into the reaction chamber **122**, according to an exemplary embodiment. The carrier gas supply manifold **320** can contain purge ports used to evacuate gases from their components, in an exemplary embodiment. The example hydrogen/deuterium gas supply manifold **310** of FIG. 3B and carrier gas supply manifold **320** of FIG. 3C, can be built, according to one embodiment, to be sufficiently rugged to contain the high pressure gases, and, for example, can be preferably constructed with high pressure, seamless ¼ inch, stainless steel tubing, or the like. The inner diameter of the exemplary tubing can be approximately ⅛ inch in an embodiment.

FIG. 3D depicts an exemplary top perspective view of an exemplary reaction gas product collection manifold **330**, according to an exemplary embodiment. Depicted is an exemplary reaction gas product collection manifold **330**, including, e.g., but not limited to, an exemplary cooling chamber **302**, exemplary mechanical valve(s) **304**, exemplary electric valve(s) **314**, an exemplary separator valve with an electronic interface for matter output (EIMO) **333**, exemplary pressure sensor **315**, and exemplary reaction gas collection tank/bottle with thermal sensor **336**, according to one exemplary embodiment. In a preferred embodiment, the exemplary electronic interface or EIMO **333** can include an acoustic or other type of electronic interface, a gas detector, and/or a reaction product sensor. In an exemplary embodiment, the electronic interface or EIMO **333** can be configured to manage material output through the separator valve and facilitate estimation of quantities of reaction gas being collected. The reaction gas product collection manifold **330**, in an exemplary embodiment, can also enable reactant gas to be temporarily stored in a separate collection tank/bottle (not shown). Advantageously, the manifold **330**, in an exemplary embodiment, can contain a purge port **316** (not labeled), which can be used for this purpose. The reactant gas or reaction gases are made to flow from the collection tank **336** through the purge port and into the separate collection tank by pressure differences between the two collection tanks. Evaluation of reactant gas collected through the purge port can be performed off-line with a commercial binary gas analyzer (for reference, see, e.g., Stanford Research Systems model BGA244), whose data can also be used to verify EIMO gas measurements. Quantities of gases to be extracted also can be determined through pressure, temperature and volume calculations for gas in the collection bottle. As for the gas supply manifolds, the reaction gas product manifold **330** is designed for safety to withstand high pressure and to contain a low gas volume, in an exemplary embodiment. The reaction gas product manifold, in an exemplary embodiment, can be designed and engineered, however, to operate only when reaction gas needs to be collected, whereupon manifold valves **314** can be operated to move gas product into the collection tank/bottle, in an exemplary embodiment. The valves can be closed during other periods of operation, such as during start-up and maintenance periods, in an exemplary embodiment. The reaction gas product collection manifold may be preferably constructed with exemplary high pressure ¼ inch stainless steel pipe with ¼ inch inner diameter, in an embodiment.

FIG. 3E depicts an exemplary top perspective view of an exemplary gas measurement and evacuation manifold **340**, according to an exemplary embodiment. Depicted is an exemplary gas measurement and evacuation manifold **340**,

including, e.g., but not limited to, exemplary cooling chamber **302**, exemplary mechanical valve(s) **304**, exemplary electric valve(s) **314**, exemplary measurement tank with thermal sensor **343**, and exemplary pressure sensor(s) (not labeled) in one embodiment. Also, as an enhanced safety measure, an exemplary separate safety tank/bottle **1112** can be provided for transfer through an exemplary safety valve rupture disc **1132** of gas from the exemplary reaction chamber **122** of FIGS. 1A and 2A, if high pressure limits are exceeded, in an exemplary embodiment. The gas measurement and evacuation manifold **340** can enable the apparatus, including the reaction chamber **122** and key portions of other manifolds, to be evacuated during maintenance periods, in an exemplary embodiment. Advantageously, the manifold **340** can contain an exemplary purge port(s) **316** for this purpose, in one exemplary embodiment. According to another aspect of the present disclosure, the measurement tank or bottle **343** has a known volume to enable quantities of extracted gas to be determined through pressure, temperature and volume calculations before the gas is evacuated from the system. The exemplary reaction gas measurement and evacuation manifold **340** is designed to withstand high pressure and to contain a low gas volume, in an exemplary embodiment. The manifold **340** can also be designed to withstand high vacuums impressed on the system during maintenance periods. The manifold **340** can be constructed with high pressure ¼ inch stainless steel pipe with ¼ inch inner diameter, in an exemplary embodiment.

Referring now to FIGS. 4A-B, the preferred embodiment of the modular electronic control circuit (ECC) or subsystem **400** can include robust off-the-shelf electronic and electrical components, to include a special-purpose computer and/or monitor, microprocessor or microcontroller **401** with control software, long-term data storage unit(s), anode-to-cathode voltage/current supply **402**, anode heater supply **403**, gas valve power supply(ies) **404**, nuclear radiation sensor electronics, starter/initiator/microwave electronics **405**, heat exchanger electronics **406** and uninterruptible power supply (UPS) **407**. The ECC can communicate with a wide number of electronic components, and can include temperature and pressure sensors for calculation of gas quantities to inject into the reaction chamber, according to an exemplary embodiment. The basic functionality of the modular ECC was described earlier in U.S. Pat. No. 6,248, 221 B 1, the content of which is incorporated herein by reference in its entirety, and will not be repeated here.

According to an exemplary embodiment, the ECC's exemplary hardware components can include a single-board computer containing a processor chip with accessory and supporting circuitry and relay boards. In an exemplary embodiment, the ECC can include, e.g., but not be limited to, a controller area network (CAN) transceiver chip and port **408**; a universal serial bus (USB) port; an RS 232 serial port; general-purpose input and output(s) (GPIO) **409**; an ethernet physical (PHY) interface and port connector; a VGA output display port (or SVGA, XGA, or HDMI) and connector; an external memory; an analog input extender board; an analog output extender board; and/or control area network containing digital output modules, digital input modules, analog output modules, analog input modules; a power supply; a process power supply; a bipolar (±), variable, high voltage and current power supply **402**; an isolated power supply; equipment power supplies; a 12 volt DC power supply; an isolated 12 volt DC power supply; a 24 volt DC power supply, an isolated 24 volt DC power supply; a video screen; a display screen; an output device; an input device; a keyboard; a touch display; or a relay board.

25

According to other exemplary embodiments, various other input devices, output devices, sensor(s), accelerometer(s), pressure sensor(s), touchscreens, communication network subsystems, wireless communication, and other components can be integrated into the system.

According to an exemplary embodiment, the ECC's exemplary software components for process control and optimization can include, e.g., but not be limited to: a multiple-input and multiple-output (MIMO) control process; a MIMO control algorithm; a proportional-integral-derivative (PID) control algorithm; a feedback controller; a Pulsed Chamber Pressurizer algorithm; a least means square (LMS) algorithm for optimization; a Gradient Search by Steepest Descent algorithm; Box-Jenkins algorithm to "system ID" linear portions of the process; a Hammerstein-Wiener System estimator to "system ID" nonlinear portions of the process; a Radial Basis Network for system modeling to "system ID" linear and/or nonlinear portions of the process; a Principal Component Analysis (PCA) system to facilitate the system modeling process; and/or an Embedded Linux or other Operating System upon which the aforesaid process control and optimization software can execute.

According to an exemplary embodiment, the Pulsed Chamber Pressurizer can include, e.g., but not be limited to, a software algorithm that can control input gas valves such that a gas chamber is methodically charged with a proportion of gas(es) to a specified setpoint(s), e.g., even if the differential pressure from the input gas supply line(s) to the reaction chamber is much greater than the reaction chamber pressure. "Charge" (in the context of gases), in an exemplary embodiment, is the accomplishing of the correct pressurization and proportion of gases while compensating for temperature change during pressurization. Alternative embodiments can be constructed from hardware using, e.g., but not limited to, a field programmable gate array (FPGA), an application specific integrated circuit (ASIC), and/or other technology. For clarification, correct pressurization is a subset of charging. "System ID" means to model, discover and identify system process characteristics under a broad range of input conditions and to create a mathematic model that accurately describes the same characteristics, where the same model can be implemented into control system software in the form of parameterization, in an exemplary embodiment.

FIG. 4A depicts an exemplary circuit architecture 400 for the ECC, according to one exemplary embodiment. One unique aspect, in one exemplary embodiment of the ECC can include, e.g., but not be limited to, real-time control of the reaction process through combined independent and interdependent application of control elements (processor outputs) indicated above: the voltage/current source between the anode and cathode (represented by 402) providing a potential difference between the anode and cathode; the anode heater 403; the heat exchanger (represented by 406); and the microwave source 405, all of which facilitate additional transport of gas between the anode and cathode and each supporting diffusion of gas into the reaction material, in one exemplary embodiment. For clarification, the exemplary process, in one exemplary embodiment, may be enabled through an exemplary single control element, multiple elements or through a combination of all elements acting synergistically.

According to another important aspect of the present disclosure, the exemplary ECC can, e.g., but not be limited to, provide, monitor, analyze and otherwise control the necessary and desired sequence of operational steps during set-up and initial steps of reactor operation, in one exem-

plary embodiment. (See exemplary numbered items in the process and software architecture depicted in FIG. 4B). Advantageously, in an exemplary embodiment, the possible amount of energy that can be produced through chemical reactions in the initial exemplary feature steps (#1) of reactor operation can be mathematically determined from known quantities and physical characteristics (e.g., mass and volume) of the reactor and cathode material, as well as included oxygen and hydrogen, for example. This excess amount of chemical energy produced during the initial steps of reactor operation, as determined by real-time pressure and temperature measurements, can then be compared against the expected energy (i.e., chemical and nuclear), in an exemplary embodiment. In an exemplary embodiment, additional sequences of carrier gas/vacuum purging steps can be performed as needed to limit the amount of aforesaid chemical energy produced.

According to another important aspect of the present disclosure, during both exemplary initial and exemplary continuous steps of exemplary reactor operation (#2/3), the ECC can provide exemplary precise control of gas pressure and proportion of the mixture of gases in the reactor, in one exemplary embodiment. This can be accomplished with the exemplary Pulsed Chamber Pressurizer algorithm applied to charging the mixing chamber and to charging the reaction chamber, in one exemplary embodiment. Due to the high amount of pressure in the gas supply and relatively low pressure and quantity of gas for the mixing chamber and reaction chamber, conventional control of gas flow is not sufficient for optimum chamber pressure and mixture precision. The gas valves can, therefore, be pulsed open for time intervals less than or equal to the valve's maximum open and close time, whereby the amount of gas in each pulse is small relative to the amount of gas required in the chamber, and the amount of gas flow is determined by the number of pulses, in one embodiment. The amount of each type of gas and the corresponding number of pulses required can be calculated algorithmically according to measured inputs, including, in one exemplary embodiment, supply tank pressure, reaction chamber volume, pressure and temperature, mixing chamber volume, pressure and temperature, and the user-determined chamber pressure and temperature setpoints, in one embodiment. In an exemplary embodiment, volumes to be loaded with deuterium and/or hydrogen gas (reference "A Theoretical Model for Low-Energy Nuclear Reactions in a Solid Matrix" by K. P. Sinha, 1999) can include the annular space around the anode and within the cathode material for which the ECC can calculate and limit loading objectively to less than a maximum allowed quantity of gas for each unit volume of reaction material (e.g., million atoms).

According to another important aspect of the disclosure, the microwave source can operate by irradiating the reaction volume (cavity) with electromagnetic radiation by methods known to those skilled in the art of microwave plasma generators, in one exemplary embodiment. See, for example, "The Large Volume Microwave Plasma Generator: A New Tool for Research and Industrial Processing," by R. G. Bosisio et al. and "Microwave Discharges: Generation and Diagnostics," by Yu. A. Lebedev. In an exemplary embodiment, the antenna for the microwave source causes the cavity between the anode and cathode to operate at a commonly used microwave frequency, e.g., 2.45 GHz, to create polarized movement of the deuterium, hydrogen and argon carrier gas mixture, thereby increasing the electron-gas molecule collision frequency. According to an aspect of the disclosure, the improved ECC can include a robust loop

27

28

or other microwave antenna that, instead of being exposed, advantageously can be encased in a ceramic tube to provide mechanical stability and otherwise protect the antenna, in one exemplary embodiment. According to another aspect of the present disclosure, the coax cable for the antenna can be connected or coupled to the antenna, and the antenna and the ceramic tube can be made as an integrated component for ease of repair or replacement during maintenance, in one exemplary embodiment.

Another unique aspect of the exemplary ECC is the example capability of continuous data collection and logging (#4) in one exemplary embodiment. Data logging can provide continuous sampling of system sensors at an example constant sampling rate, as determined by engineering and scientific personnel prior to and/or during system operation. Data logging can be automatic and processor-based, in one exemplary embodiment. In one exemplary embodiment, system software can detect a single sensory event or any multiple combinations of sensor events occurring within the logged data or in real time. System control parameters can be determined empirically from a computer analysis of the logged data and from knowledge of significant system events extracted therefrom and enabled by System ID, in one exemplary embodiment. The information contained in the same logged events can be used to improve and optimize reactor operation through adjustment of control parameters and internal algorithm parameters as well as thresholds for fail-safe operation enabled by adaptive processes, in one exemplary embodiment. This capability can also recognize patterns amongst the sensory inputs that can indicate any of a number of system conditions, including hazard to persons or equipment, and can control system component failure whether partial, intermittent or full and system operational inefficiencies, in one exemplary embodiment.

Exemplary Electronic Control Circuit Tasks/Functions in an exemplary embodiment include:

1. Provide automated system pre-startup sequence;
2. Provide automated charging of mixing chamber and reaction chamber;
3. Control reaction chamber dynamics during startup and/or operation;
4. Provide automatic fail-safe shut down procedures;
5. Provide long term application and/or automatic adjustment of control system output signals to the process to facilitate the empirical discovery of the signal levels required to initiate, control and/or maintain a desired reaction and to otherwise enable controllability and repeatability;
6. Provide process measurements for real-time human operator observation;
7. Enable human interface with the process through video screens where an operator or engineer can initiate and/or terminate a process, and/or adjust specified control system parameters to change the speed, nature and intensity of a process;
8. Provide detection of significant system events and/or combinations thereof, measured from sensory input field devices, and/or further provide a corresponding threshold table upon which the control system automatically responds;
9. Provide a collection of data from one or more, or all sensory input devices sampled and stored in an exemplary static, external device connected or coupled through, e.g., the processor board USB port;

10. Provide adaptive algorithms for system self adjustment and/or system identification algorithms for modeling of system characteristics; and/or
11. Measure the excess heat as a function of time.

In one embodiment, the input valve(s) can be pulsed such that no temperature-compensated pressure setpoint is overshot by more than a specified value (e.g., 5 psi) and such that the specified charging process completes in less than a time setpoint where pressure(s) are within a differential psi of the temperature-compensated setpoint(s) according to one exemplary embodiment. Selection of gas(es), the proportion thereof, and pressurization setpoint(s) are engineer specified, in one embodiment. Furthermore, the charge speed versus overshoot risk can be operator controlled through the use of engineer accessible control parameters, in one embodiment. A threshold table can be built in software containing a list of actions, some initiated by interrupt, corresponding to a list of events characterized by sensory device input, or combinations thereof, where the system fails to maintain or timely achieve thresholds, in one embodiment.

According to an exemplary embodiment, an electronic control system can include a processor (e.g., FPGA, PLC, embedded CPU, ARM-based controller, microcontroller, ASIC, etc.), wherein the reaction contained in the reaction chamber can be controlled by a computer implemented control algorithm and control algorithm parameters are determined adaptively through an iterative or empirical process. According to an exemplary embodiment, the activity in the reaction chamber can have some nonlinearity, thus, an embodiment can include a learning algorithm configured to optimally adjust parameters. The system can be adaptive, in an exemplary embodiment. The system can use feedback control and be adaptive. It can also, for example, be connected with, or coupled to, and provide signaling to mechanical steam controllers and pressure valves for different possible power applications, see, e.g., but not limited to, FIG. 5.

Assembly and Operation

These details can enable one skilled in the art to assemble the apparatus' custom-designed and off-the-shelf parts without an undue amount of additional research into system design. Many technology companies already have much of the needed electronic equipment. The manufacturer of consolidated cathode reaction material, as example, may use the apparatus to quantify the ability of manufactured cathodes to absorb deuterium or hydrogen gas. The system would enable the reaction material initially to be subjected to high vacuum and subsequently to measured quantities of one gas or both. Temperature and pressure measurements recorded during steps of the process can be used to quantify amounts of gas able to be absorbed, and these data then used to improve on reaction material design and for quality control of the cathode manufacturing process. The operator may use the apparatus to investigate the possible amount of energy produced through chemical reactions. The nuclear scientist investigating types of reactions in the cathode may use the apparatus to quantify the amounts of reaction products produced. The system would enable small samples of reaction product gases to be extracted from the reaction gas product manifold and subjected to further measurements and study. The space power engineer may use the apparatus to quantify any amount of thrust that could be provided from the heat exchanger.

Assembly.

A new cathode is inserted into the reaction chamber and a new heater is inserted into the anode. The header is mated to internal support insulators and connections are made to header feedthroughs from the anode, anode heater, microwave antenna and thermocouple temperature sensor. The completed header assembly is lowered into the reactor and sealed with metallic seals. The reaction chamber is fitted into the heat exchanger and sealed. The four gas manifolds are mated with the reaction chamber. Electrical connections are made with the header feedthroughs and manifold temperature and pressure sensors and valves. After mechanical and electrical assembly, the system's software is loaded into its computer or microprocessor and exercised to demonstrate that all operations are appropriately controlled. The computer/data acquisition system would be set to continuously (e.g., each second) record data from all voltage and current sources, temperature and pressure sensors, and valve settings.

Preparation for Operation.

The sequence of operations would begin with all valves closed. The system is pressurized and valves cycled to ensure that leaks do not occur from the reaction chamber, header and manifolds. The reactor would then be subjected to a sequence of vacuum and high temperature cycles. Carrier gas would be provided to help remove oxygen. The system under vacuum is allowed to bake until no further pressure changes occur. At this time, the cathode is assumed to be depleted of absorbed gases. Vacuum pumping is valved off. The natural pressure rise in the system is documented versus time.

Fueling and Loading.

The sequence of operations from this point would vary according to operational objectives. In general, a small increment of deuterium would be provided first to the reaction chamber before any hydrogen. Pressure inside the reaction chamber would be monitored to determine loading of the cathode with gas; and, loading steps would be repeated as needed. After sufficient loading, the reaction chamber would be subjected to high temperature from the anode, and to high voltage potential on the anode. The heat exchanger would be cycled briefly to cool the outer surface of the reaction chamber. These loading steps would be repeated as additional deuterium and/or hydrogen is loaded into the cathode. Data recorded by the computer would be used to determine the efficiency with which the cathode is loaded and the amount of any energy produced.

Operation and Mature Operation.

The fueled reaction chamber is now ready for operation. The heat exchanger and other elements of the cooling system, anode heater and anode voltage are engaged. The cathode remains at ground potential. The reaction chamber pressure and temperature are monitored to ensure that operating limits are not exceeded. Loading steps would be repeated as additional deuterium and/or hydrogen is required to be loaded into the cathode. Microwave excitation can be applied to improve ionization and potentially excite the cathode lattice structures. Data recorded by the computer is used to determine the efficiency with which the cathode is loaded and the amount of energy produced during operation. Electronic computer control operates the system according to planned set points. After some period of operation, temperature in the reaction chamber can be expected to decrease as gas is consumed and reaction products increase. Computer control can reestablish the operating point by quickly pulsing the gas valves to admit the necessary quantity of deuterium/hydrogen. Reaction gas product is removed after a sufficiently long period of operation by operating valves in the collection gas manifold. The exhaust gas can then be further processed as desired.

Quenching and Shutdown.

Input energy (anode heat, anode-to-cathode current and microwave excitation) is terminated when it is desired to stop the reactor. The reactor will relax to an energy neutral state and the cooling system is used to reduce the temperature to ambient. Inert carrier gas can be injected to quench the system. When this is complete, the vacuum system can be used to empty the reactor in preparation for maintenance or the next operational cycle.

If demonstrated to produce energy efficiently, the present disclosure has potential application in many situations that require a long-term, continuous source of heat energy with minimal environmental and safety concerns, inexpensive and small-volume fuel requirements, and simplified operational procedures with little necessary monitoring. In many possible applications greatly reduced initial capital costs can be expected. Following are a few exemplary, but non-limiting examples of its potential applications:

Primary use can include replacement for existing heat sources in conventional and nuclear power plants. Example advantages can include lack of fossil fuel emissions and fossil fuel requirements. In nuclear systems, example advantages can include lack of radioactive waste, minimal shielding requirements and standoff distances, and elimination of large onsite inventory of high-level radioactive materials (i.e., avoidance or elimination of meltdown risk of Fukushima, Chernobyl, Three Mile Island and other potentials). Arrays of supplemental thermoelectric generator elements can be used to create energy from heat. Further the heat source can be used for desalinization of water, in exemplary embodiments.

Potential use can include distributed power production systems. Localized (such as, e.g., but not limited to, city, neighborhood, individual building) power plants can use these embodiments. Example advantages over traditional centralized generation and distribution can include, but are not limited to, cost savings from lack of high maintenance distribution networks, improved efficiency from lack of transmission losses, and reduced susceptibility to mass failures (from, e.g., but not limited to, natural disasters, war, terrorism, electromagnetic pulse (EMP) disturbance, coronal mass ejection (CME), etc.).

Many nations maintain outposts and research facilities in polar and/or other remote and/or other power accessibility poor regions. These embodiments can potentially support these needs and greatly reduce associated environmental concerns. By comparison, conventional nuclear sources are environmentally unpalatable and conventional fossil fuel systems can have enormous logistical problems in these application.

Deep space operations have a critical need for long-lived power sources, as may be provided by embodiments in this disclosure. Sufficient conventional fuels cannot be conventionally carried along. Solar power becomes less useful as the distance from the sun increases. A nuclear isotope thermoelectric generator (i.e., a current solution) can be expensive and can present unusual risks in launch. The embodiment in this disclosure can require minimal and lightweight fuel requirements and potentially replace isotope cores in thermoelectric generator systems.

The system in this disclosure can potentially be used in modular designs where some modular systems are impractical. Accordingly, the exemplary system can include the use of multiple, identical and aggregated and/or coordinated systems, plus there can be an environmental benefit. An exemplary embodiment can support distributed energy systems and can support an exemplary neighborhood system, which by its close locality, can avoid transmission losses.

A system derived from this disclosure can potentially be very useful in desalinization and/or other water producing facilities worldwide as a primary heat source, or for providing electricity. Such facilities will likely be in high demand based on current population projections.

It is possible that the system can be adapted to meet propulsion requirements for both land-based and seagoing transport vehicles and become an alternative to conventional nuclear technology in subsurface vehicles.

Industrial operations, such as chemical processing, synthesizing, and refining, along with many manufacturing facilities, have requirements for sources of heat that do not have the safety concerns of conventional burning of fossil fuel. Similar considerations exist in facets of mining industries. Systems derived from this disclosure could reduce fire and explosion hazards and minimize contamination, unlike conventional sources.

Smaller power sources can be very useful in regions of the world that have little or no modern infrastructure. According to other exemplary embodiments, multiple uses can range from small direct thermoelectric generators to sterilization facilities for rural or temporary clinics.

Various other possible applications of energy and heating can be used, as will be apparent to those skilled in the relevant art, according to an exemplary embodiment.

SUMMARY OF THE PREFERRED EXEMPLARY EMBODIMENTS

According to a preferred embodiment, the gaseous electrolysis apparatus comprises a cooled header with at least one electrical connector or coupling; a heat exchanger configured to remove heat from a surface of the reaction chamber; a gas handling system mechanically coupled to the reaction chamber; and, an electronic control circuit electrically connected to the header and gas handling system.

According to a preferred embodiment, the said header, as shown in FIGS. 1D and 6B, comprises a cooling apparatus; cooling manifold or water jacket and at least one feedthrough to a header cooling manifold proximate to said header.

According to a preferred embodiment, at least one feedthrough, as shown in FIGS. 1B-C and 6C, comprises a pressure side oriented towards the inside of the reaction chamber, is welded into a thermal plate or comprises a threaded coupling, and extends beyond the header for connection with the electronic control circuit.

According to a preferred embodiment, the header, as shown in FIGS. 1B-C and 2A, comprises an anode connection, anode heater wire connections, microwave antenna connection, thermal sensor connection, microwave loop antenna, and insulator(s) configured to electrically isolate, minimize the volume where gas resides, and provide mechanical support for components within the reaction chamber.

According to a preferred embodiment, the gaseous electrolysis apparatus further comprises a modular, removable anode, as shown in FIGS. 2A-B and 6D. The edges of the anode facing the cathode are tapered or curved to help prevent high voltage breakdown between the anode and cathode.

According to a preferred embodiment, the gaseous electrolysis apparatus further comprises a modular, removable, hollow-shaped, cylindrical cathode, as shown in FIGS. 2A-B and 6E, with a central cavity configured to receive the

anode, encased by an outer metal supporting sleeve, and bounded at its base and top with insulator endcaps. The cathode is co-disposed about the anode within the reaction chamber and the cathode and anode are coaxial. The edges of the reaction material part facing the anode are tapered or curved to help prevent high voltage breakdown between the anode and cathode.

According to a preferred embodiment, the heat exchanger, as shown in FIG. 2C, is modular and comprises a relatively low volume flash boiler engineered to provide a mist of water or other coolant to the outer surface of the reaction chamber, a plurality of spray nozzles to cool at least one portion of the reaction chamber, at least one steam pressure port, and at least one thruster port configured to provide pressure output. The heat exchanger is co-disposed around the reaction chamber.

According to a preferred embodiment, the gas handling system, as shown in FIGS. 3A-E, comprises four separate gas manifolds that can control gas flow while minimizing gas volume external to the reaction chamber: a hydrogen/deuterium gas supply manifold; an inert carrier gas manifold; a reaction gas product collection manifold; and a gas measurement and evacuation manifold.

According to a preferred embodiment, the gas manifolds, as shown in FIGS. 3A-E, comprise a cooling chamber or water jacket to provide cooling for gas tubing and pipes connected or coupled to the reaction chamber, gas compatible valves, pressure and temperature sensors connected to the electronic control circuit/subsystem, tanks or containers whose known volume enables small quantities of gas to be determined by calculating pressure, temperature and volume before gas is transferred into or out of the reaction chamber, and purge ports which can be used to evacuate gases manually.

According to a preferred embodiment, the reaction gas product collection manifold comprises a container configured to temporarily store reactant gas and periodically permit extraction from the container and an acoustic sensor subsystem or other type of electronic interface to facilitate estimation of reaction product gas quantity, as shown in FIGS. 3D and 6F.

According to a preferred embodiment, the electronic control circuit (ECC), as shown in FIGS. 4-5, comprises off-the-shelf electronic and electrical components, to include a special-purposed computer and/or monitor, microprocessor or microcontroller with control software; long-term data storage unit(s); special-purposed anode-to-cathode voltage/current supply; anode heater supply; gas valve power supplies; nuclear radiation sensor electronics; microwave starter/initiator electronics; heat exchanger electronics; and uninterruptible power supply (UPS).

Other modifications and variations to the disclosure will be apparent to those skilled in the art from the foregoing disclosure and teachings. Thus, while only certain embodiments of the disclosure have been specifically described herein, it will be apparent that numerous modifications may be made thereto without departing from the spirit and scope of the disclosure.

Therefore, we claim:

1. A modular gaseous electrolysis apparatus, wherein an electrolyte is gas, the gaseous electrolysis apparatus comprising:

an actively-cooled header module with at least one electrical connector or coupling, and

wherein said actively-cooled header module is configured to be one of:

opened/removed, and

closed/installed;

a heat exchanger module configured to:

remove heat from a surface of a reaction chamber module; and

facilitate and control a thermal gradient through a removable cathode module, and a wall of the reaction chamber module, wherein said heat exchanger module is separate from the removable cathode module, and wherein the reaction chamber module is configured to receive the gas electrolyte;

wherein said heat exchanger module is capable of assembly, and disassembly, and

wherein the gaseous electrolysis apparatus comprises at least one or more of:

wherein the gaseous electrolysis apparatus is configured to include hermetic seals to maintain integrity in an elevated pressure and temperature environment;

a flash boiler is configured to provide a mist of water or other coolant to the outer surface of the reaction chamber module; or

a plurality of spray nozzles is configured to cool at least one portion of the reaction chamber module and to facilitate and control thermal diffusion through the removable cathode module; and

at least one or more of:

at least one steam pressure port; or

at least one thruster port configured to provide pressure output;

a gas handling system, configured to provide the gas electrolyte, said gas handling system mechanically coupled to the reaction chamber module and separate from the reaction chamber module,

wherein said gas handling system comprises at least one or more of:

a measurement container configured to temporarily store reactant gas and periodically permit extraction of the reactant gas from said measurement container; or

a subsystem comprising an acoustic sensor or other type of electronic interface, configured to facilitate estimation of quantities of reactant gas;

an electronic control circuit module electrically coupled or connected to said actively-cooled header module and said gas handling system and configured to electronically control said gas handling system;

a modular, removable anode module comprising:

an electric heater disposed within the modular, removable anode module;

wherein said actively-cooled header module, said heat exchanger module, the removable cathode module, said modular, removable anode module, said electric heater, said gas handling system, and said electronic control circuit module are removably and mechanically coupled to the reaction chamber module; and

wherein said gas handling system comprises:

a gas manifold module that controls gas flow external to the reaction chamber module,

wherein said gas manifold module comprises:

a reaction gas product collection manifold module, mechanically coupled to said reaction chamber module, and coupled to a reaction gas product collector and further comprising:

at least one mechanical valve;

at least one electronically controlled valve;

at least one separator valve with an electronic interface for matter output (EIMO);

at least one pressure sensor;

at least one exemplary reaction gas collection tank or bottle comprising a thermal sensor; and

wherein said at least one electronic interface for matter output (EIMO) comprises at least one or more of:

the at least one acoustic sensor or other type of electronic interface;

at least one gas detector, or

at least one reaction product sensor; and

wherein said at least one electronic interface for matter output (EIMO) comprises being configured to at least one or more of:

manage material output through said at least one separator valve; or

facilitate estimation of quantities of reaction gas being collected.

2. The gaseous electrolysis apparatus according to claim 1, wherein said actively-cooled header module is configured to enable the gaseous electrolysis apparatus to be operated for a period of time between maintenance periods, and wherein said actively-cooled header module comprises at least one or more of:

a physically extended cooling manifold or water jacket, to improve thermal efficiency; or

at least one feedthrough to a header module cooling manifold proximate to said actively-cooled header module.

3. The gaseous electrolysis apparatus according to claim 2, comprising said at least one feedthrough configured and constructed with at least one conductor surrounded by at least one insulating material to maintain integrity of electronics fed therethrough during a period of at least one or more of: a variable duration, or a long duration, of elevated pressure and temperature environment, and

wherein said at least one feedthrough comprises at least one or more of:

is welded into a thermal plate;

wherein one end of said feedthrough extends beyond said actively-cooled header module for connection with the electronic control circuit module; or

wherein said at least one feedthrough comprises a threaded coupling.

4. The gaseous electrolysis apparatus according to claim 1, wherein said actively-cooled header module comprises:

a gasket to seal the actively-cooled header module to a top of a body of the reaction chamber module;

at least one anode module connection;

at least one anode module heater wire connection configured such that a heater is configured to raise a temperature of reaction material in the removable cathode module;

at least one thermal sensor connection;

at least one ceramic-encased microwave loop antenna configured to:

facilitate transport of the gas electrolyte between an anode module and the removable cathode module; and

support diffusion of the gas electrolyte into a reaction material; and

at least one insulator configured to at least one or more of:

electrically isolate;

minimize a volume where the gas electrolyte resides; or

provide mechanical support for components of the heater module within the reaction chamber module.

5. The gaseous electrolysis apparatus according to claim 1, wherein the

modular, removable anode module further comprises

wherein edges at the ends of said modular, removable anode module facing the removable cathode module are tapered or curved to help prevent high voltage breakdown between the modular, removable anode module and the removable cathode module.

6. The gaseous electrolysis apparatus according to claim 1, wherein the removable cathode module comprises:

a modular, removable, hollow-shaped, cylindrical removable cathode module, electrically coupled, connected, or grounded to the reaction chamber module with a central cavity configured to receive an anode module;

wherein the removable cathode module is-encased by an outer metal supporting sleeve;

wherein the removable cathode module is bounded at a base and at a top of the removable cathode module with at least one insulator endcap; and

wherein edges at the ends of a reaction material part of the removable cathode module facing the anode module are at least one of tapered or curved to help prevent high voltage breakdown.

7. The gaseous electrolysis apparatus according to claim 1, wherein said heat exchanger module comprises at least a portion of a space surrounding the reaction chamber module, and

is co-disposed around the reaction chamber module.

8. The gaseous electrolysis apparatus according to claim 1, wherein said gas handling system comprises: four (4) separate gas manifold modules, coupled to said reaction chamber module, said four separate gas manifold modules are configured to control gas flow external to the reaction chamber module, while serving to minimize gas volume external to said reaction chamber module,

wherein said four separate gas manifold modules comprise:

a hydrogen/deuterium gas supply manifold module, mechanically coupled to said reaction chamber module, and coupled to a hydrogen/deuterium supply gas supply source;

an inert carrier gas manifold module, mechanically coupled to said reaction chamber module, and coupled to an inert gas supply source;

a reaction gas product collection manifold module, mechanically coupled to said reaction chamber module, and coupled to a reaction gas product collector; and

a gas measurement and evacuation manifold module, mechanically coupled to said reaction chamber module, and coupled to a gas measurement and evacuation device.

9. The gaseous electrolysis apparatus according to claim 8, wherein each of said four separate gas manifold modules, coupled to said reaction chamber module, comprises:

a cooling chamber or water jacket to provide cooling for gas tubing and pipes connected or coupled to the reaction chamber module;

at least one normally closed, gas compatible valve, at least one pressure sensor and at least one temperature sensor connected to, or coupled to said electronic control circuit module/subsystem; and

at least one tank or at least one container whose known volume enables small quantities of gas to be determined by calculating pressure, temperature and volume before gas is transferred into or out of the reaction chamber module.

10. The gas electrolysis apparatus according to claim 8, wherein the reaction gas product collection manifold module comprises:

the measurement container configured to temporarily store reactant gas and periodically permit extraction of the reactant gas from said measurement container; and

the subsystem comprising the acoustic sensor or the other type of electronic interface configured to facilitate estimation of quantities of the reactant gas.

11. The gaseous electrolysis apparatus according to claim 1, wherein said electronic control circuit (ECC) module comprises:

an automated special-purpose computer and display monitor, and control software;

automated gas handling system electronics; wherein said automated gas handling system electronics is coupled to, and related to associated electric valves, temperature sensors and pressure sensors;

automated anode-to-cathode voltage/current supply;

automated heater supply;

automated microwave starter or initiator electronics; and

automated heat exchanger module electronics.

12. The gaseous electrolysis apparatus according to claim 1, wherein said gas handling system comprises at least one gas manifold module, coupled to said reaction chamber module, and is configured to control gas flow, external to the reaction chamber module, while serving to minimize gas volume external to said reaction chamber module,

wherein said at least one gas manifold module comprises at least one or more of:

a hydrogen/deuterium gas supply manifold module, mechanically coupled to said reaction chamber module, and coupled to a hydrogen/deuterium supply gas supply source;

an inert carrier gas manifold module, mechanically coupled to said reaction chamber module, and coupled to an inert gas supply source;

a reaction gas product collection manifold module, mechanically coupled to said reaction chamber module, and coupled to a reaction gas product collector; or

a gas measurement and evacuation manifold module, mechanically coupled to said reaction chamber module, and coupled to a gas measurement and evacuation device.

13. The gaseous electrolysis apparatus according to claim 12, wherein said at least one gas manifold module, coupled to said reaction chamber module, comprises at least one or more of:

a cooling chamber or water jacket to provide cooling for gas tubing and pipes connected or coupled to the reaction chamber module;

at least one normally closed gas compatible valve, at least one pressure sensor and at least one temperature sensor connected to, or coupled to said electronic control circuit module/subsystem; or

at least one tank or at least one container whose known volume enables small quantities of gas to be determined by calculating pressure, temperature and volume before gas is transferred into or out of the reaction chamber module.

14. The gaseous electrolysis apparatus according to claim 1, further comprising:

wherein said modular, removable anode module comprises

wherein said modular, removable anode module comprises wherein edges at the ends of said modular, removable anode module facing the removable cathode module are tapered or curved to help prevent high voltage breakdown between said modular, removable anode module and the removable cathode module; and

wherein said electric heater module disposed within said modular, removable anode module comprises:

wherein said electric heater module is configured to raise the temperature of the reaction material in the removable cathode module, by at least one of thermal radiation or diffusion; and

wherein said electric heater module is electrically connected through a feedthrough in the header module to a power supply configured to provide power to said electric heater module.

15. The gaseous electrolysis apparatus according to claim 1, wherein the removable cathode module comprises at least one or more of:

a modular, removable, hollow-shaped, cylindrical removable cathode module, electrically coupled, connected, or grounded to the reaction chamber module with a central cavity configured to receive an anode module;

wherein the removable cathode module is encased by an outer metal supporting sleeve;

wherein the removable cathode module is bounded at a base and at a top of the removable cathode module with at least one insulator endcap; or

wherein edges at each ends of a reaction material part of the removable cathode module facing the anode module are at least one of tapered or curved away from the anode module to help prevent high voltage breakdown.

16. The gaseous electrolysis apparatus according to claim 1, wherein said heat exchanger module is modular and comprises:

a flash boiler configured to provide a mist of water or other coolant to the outer surface of the reaction chamber module; and

a plurality of spray nozzles to cool at least one portion of the reaction chamber module and to facilitate and control thermal diffusion through the removable cathode module; and one or more of

the at least one steam pressure port; or

the at least one thruster port configured to provide pressure output.

17. The gas electrolysis apparatus according to claim 1, wherein said reaction gas product collection manifold module further comprises at least one or more of:

a measurement container, coupled to said reaction gas product collection manifold module, configured to temporarily store reactant gas and periodically permit extraction of the reactant gas from said measurement container; or

a subsystem, coupled to said reaction gas product collection manifold module, comprising said acoustic interface, wherein said acoustic interface comprises an acoustic sensor configured to facilitate estimation of quantities of the reactant gas.

18. The gas electrolysis apparatus according to claim 1, wherein said reaction gas product collection manifold module further comprises:

a measurement container configured to temporarily store reactant gas and periodically permit extraction of the reactant gas from said container; and

a subsystem comprising at least one of:

an acoustic sensor, or other type of electronic interface, said subsystem configured to facilitate estimation of quantities of the reactant gas.

19. The gaseous electrolysis apparatus according to claim 1, wherein said electronic control circuit (ECC) module comprises at least one or more of:

an automated special-purpose computer and display monitor, and control software;

automated gas handling system electronics, wherein said automated gas handling system electronics is coupled to, and related to associated electric valves, temperature and pressure sensors;

automated anode-to-cathode voltage/current supply;

automated anode module heater supply;

automated microwave starter or initiator electronics; or

automated heat exchanger module electronics.

20. A modular gaseous electrolysis apparatus, wherein an electrolyte comprises gas, the gaseous electrolysis apparatus comprising:

an actively-cooled header module comprising at least one electrical connector or coupling, and

wherein said actively-cooled header module is configured to be at least one or more of:

opened,

removed,

closed, or

installed;

a heat exchanger module configured to:

remove heat from a surface of a reaction chamber module; and

facilitate and control a thermal gradient through a removable cathode module, and

a wall of the reaction chamber module,

wherein said heat exchanger module is separate from the removable cathode module, and

wherein the reaction chamber module is configured to receive the gas electrolyte;

wherein said heat exchanger module is configured to be at least one or more of assembled, or disassembled, and

wherein the gaseous electrolysis apparatus comprises at least one or more of:

wherein the gaseous electrolysis apparatus is configured to include hermetic seals to maintain integrity in an elevated pressure and temperature environment;

a flash boiler configured to provide a mist of water or other coolant to the outer surface of the reaction chamber module; or

a plurality of spray nozzles configured to at least one or more of:

cool at least one portion of the reaction chamber module;

facilitate thermal diffusion; or

control thermal diffusion through the removable cathode module; and

at least one or more of:

at least one steam pressure port; or

at least one thruster port configured to provide pressure output;

a gas handling system, configured to provide the gas electrolyte, said gas handling system mechanically coupled to the reaction chamber module and separate from the reaction chamber module,

wherein said gas handling system comprises at least one or more of:

a measurement container configured to temporarily store reactant gas;

a measurement container configured to permit extraction of the reactant gas; or a subsystem comprising an acoustic sensor or other type of electronic interface, configured to facilitate estimation of quantities of reactant gas;

an electronic control circuit module electrically coupled or connected to said actively-cooled header module and said gas handling system and configured to electronically control said gas handling system;

a modular, removable anode module comprising:

an electric heater disposed within the modular, removable anode module;

wherein said actively-cooled header module, said heat exchanger module, the removable cathode module, said modular, removable anode module, said electric heater, said gas handling system, and said electronic control circuit module are removably and mechanically coupled to the reaction chamber module; and

wherein said gas handling system comprises:

a gas manifold module that controls gas flow external to the reaction chamber module,

wherein said gas manifold module comprises:

a reaction gas product collection manifold module, mechanically coupled to the reaction chamber module, and coupled to a reaction gas product collector and further comprising:

at least one mechanical valve;

at least one electronically controlled valve;

at least one separator valve with an electronic interface for matter output (EIMO);

at least one pressure sensor;

at least one exemplary reaction gas collection tank or bottle comprising a thermal sensor; and

at least one electronic interface for matter output (EIMO) comprising at least one or more of:

the at least one acoustic sensor or the other type of electronic interface;

at least one gas detector, or

at least one reaction product sensor; and

wherein the at least one electronic interface for matter output (EIMO) comprises being configured to at least one or more of:

manage material output through said at least one separator valve; or

facilitate estimation of quantities of reaction gas being collected.

* * * * *

Printed in the United States
by Baker & Taylor Publisher Services

Printed in the United States
by Baker & Taylor Publisher Services